AF389168

Plasmas Processes Applied on Metals and Alloys

Plasmas Processes Applied on Metals and Alloys

Editors

Jian-Zhang Chen
Shih-Hang Chang

MDPI • Basel • Beijing • Wuhan • Barcelona • Belgrade • Manchester • Tokyo • Cluj • Tianjin

Editors
Jian-Zhang Chen
National Taiwan University
Taiwan

Shih-Hang Chang
National I-lan University
Taiwan

Editorial Office
MDPI
St. Alban-Anlage 66
4052 Basel, Switzerland

This is a reprint of articles from the Special Issue published online in the open access journal *Metals* (ISSN 2075-4701) (available at: https://www.mdpi.com/journal/metals/special_issues/plasmas_processes).

For citation purposes, cite each article independently as indicated on the article page online and as indicated below:

LastName, A.A.; LastName, B.B.; LastName, C.C. Article Title. *Journal Name* **Year**, *Volume Number*, Page Range.

ISBN 978-3-0365-0916-7 (Hbk)
ISBN 978-3-0365-0917-4 (PDF)

Contents

About the Editors

Jian-Zhang Chen (Professor) was born in 1973 in Taipei, Taiwan. Jian-Zhang Chen obtained his Bachelor's Degree from the Department of Mechanical Engineering in 1996 and Master's Degree from the Graduate Institute of Materials Science and Engineering in 1998, both at National Taiwan University. After two years of military service, he travelled to the Department of Electrical Engineering at Princeton University and received a Ph.D. in 2006. He joined the faculty of National Taiwan University in 2007 and was promoted to full professor in 2016. His current research interests are low-pressure plasma and atmospheric-pressure plasma material processing, wide bandgap electronic materials and devices, flexible electronics, supercapacitors, dye-sensitized solar cells, perovskite solar cells, AI-assisted plasma technology, and solar hydrogen generation.

Shih-Hang Chang (Professor) received his M.S. and Ph.D. degrees in materials science and engineering from National Taiwan University, Taiwan, in 2003 and 2007, respectively. He is currently a Professor with the Department of Chemical and Materials Engineering, National I-Lan University, Taiwan. He has published about 60 refereed professional research papers. His research interests include atmospheric pressure plasma, surface modifications, organic and polymer coatings, shape memory alloys, high damping materials, biomaterials, and microbial fuel cells.

Editorial

Plasmas Processes Applied on Metals and Alloys

Jian-Zhang Chen [1],* and Shih-Hang Chang [2],*

[1] Graduate Institute of Applied Mechanics, National Taiwan University, Taipei 10617, Taiwan
[2] Department of Chemical and Materials Engineering, National I-Lan University, I-Lan 26047, Taiwan
* Correspondence: jchen@ntu.edu.tw (J.-Z.C.); shchang@niu.edu.tw (S.-H.C.)

Received: 19 February 2020; Accepted: 6 March 2020; Published: 7 March 2020

1. Introduction and Scope

Plasma technology has been extensively used for many applications, such as plasma etching, thin-film deposition, and surface modification. A low-pressure plasma is easier to generate, more stable, and operated in a cleaner vacuum environment; therefore, it finds widespread use in industry. Atmospheric-pressure plasma is operated at a regular pressure without using vacuum chambers and pumps; this could lower cost, reduce maintenance effort, and enable easier integration into a reel-to-reel manufacturing process. Recently, many research and development groups have, therefore, focused on atmospheric-pressure plasma technology.

This Special Issue focuses on recent advances in plasma technology and its application to metals, alloys, and related materials; surface modification, material syntheses, cutting and surface coatings are performed using low-pressure plasma or atmospheric-pressure plasma.

2. Contributions

Nine research contributions are published in this Special Issue. The application of low-temperature air-plasma treatment on an organic-inorganic halide perovskite film was investigated using diffuse coplanar surface barrier discharge (DCSBD) at 70 °C [1]. An aluminum surface was treated and the wettability distribution was experimentally investigated using an atmospheric-pressure remote plasma [2]. Spark plasma sintering and a selective laser melting technique were used in combination to fabricate TiO_2/Ag ceramics and Ti_6Al_4V-TiO_2/Ag composites. This technique provides a feasible route to add ceramic reinforcement to 3D printed metals and alloys [3]. An atmospheric-pressure pulsed-arc plasma jet was used for the nitridation of steel. A nitrogen/hydrogen mixed gas was used and the nitrogen dose was successfully controlled [4]. A self-lubricating plasma electrolytic oxidation-polytetrafluoroethylene (PEO-PTFE) composite was coated on pure titanium. The PEO-PTFE deposited by this method shows excellent tribological properties with a low friction coefficient and wear rate [5]. A DC-pulse atmospheric-pressure plasma jet (APPJ) was used for rapidly synthesizing Pt-SnO_x nanomaterials that were then used as the counter electrodes of dye-sensitized solar cells (DSSCs). The DSSC performance can be significantly improved with only 5 s APPJ processing. The DC-pulse APPJ was demonstrated to be an efficient tool for the rapid synthesis of Pt-SnO_x nanomaterials [6]. The mechanical and microstructural features of plasma-cut steel were investigated. Further, the mechanical properties, microstructure, hardness, and residual stresses were compared and discussed [7]. The crystal structures of GaN nanodots produced by nitrogen plasma treatment on Ga metal droplets were investigated. The formation of a thin SiN_x layer could inhibit the phase transformation of GaN nanodots from a zinc-blende phase to a wurtzite phase [8]. Al_2O_3 coatings were prepared on Ti-45Al-8.5Nb alloys via cathodic plasma electrolysis deposition. The results suggested that the solution surface tension influences the average diameter of the hydrogen bubbles formed on the cathode surface during this process [9].

3. Conclusions and Outlook

The contributions of this Special Issue include a wide scope of plasma technologies applied to materials. Plasma is a versatile tool that can be applied in many types of material processing. New materials processing applications of plasmas and new plasma technologies are still being developed rapidly. As Guest Editors, we hope that this Special Issue can contribute new knowledge to the plasma materials research society.

Conflicts of Interest: The authors declare no conflict of interest.

References

1. Shekargoftar, M.; Jurmanová, J.; Homola, T. A Study on the Effect of Ambient Air Plasma Treatment on the Properties of Methylammonium Lead Halide Perovskite Films. *Metals* **2019**, *9*, 991. [CrossRef]
2. Muñoz, J.; Rincón, R.; Calzada, M.D. Spatial Distribution of Wettability in Aluminum Surfaces Treated with an Atmospheric-Pressure Remote-Plasma. *Metals* **2019**, *9*, 937. [CrossRef]
3. Rahmani, R.; Rosenberg, M.; Ivask, A.; Kollo, L. Comparison of Mechanical and Antibacterial Properties of TiO$_2$/Ag Ceramics and Ti6Al4V-TiO$_2$/Ag Composite Materials Using Combined SLM-SPS Techniques. *Metals* **2019**, *9*, 874. [CrossRef]
4. Ichiki, R.; Kono, M.; Kanbara, Y.; Okada, T.; Onomoto, T.; Tachibana, K.; Furuki, T.; Kanazawa, S. Controlling Nitrogen Dose Amount in Atmospheric-Pressure Plasma Jet Nitriding. *Metals* **2019**, *9*, 714. [CrossRef]
5. Ren, L.; Wang, T.; Chen, Z.; Li, Y.; Qian, L. Self-Lubricating PEO-PTFE Composite Coating on Titanium. *Metals* **2019**, *9*, 170. [CrossRef]
6. Lee, C.C.; Huang, T.M.; Cheng, I.; Hsu, C.C.; Chen, J.Z. Time Evolution Characterization of Atmospheric-Pressure Plasma Jet (APPJ)-Synthesized Pt-SnO$_x$ Catalysts. *Metals* **2018**, *8*, 690. [CrossRef]
7. Aldazabal, J.; Martín-Meizoso, A.; Klimpel, A.; Bannister, A.; Cicero, S. Mechanical and Microstructural Features of Plasma Cut Edges in a 15 mm Thick S460M Steel Plate. *Metals* **2018**, *8*, 447. [CrossRef]
8. Su, Y.Z.; Yu, I.S. Crystal Structures of GaN Nanodots by Nitrogen Plasma Treatment on Ga Metal Droplets. *Metals* **2018**, *8*, 419. [CrossRef]
9. Yang, X.; Jiang, Z.; Ding, X.; Hao, G.; Liang, Y.; Lin, J. Influence of Solvent and Electrical Voltage on Cathode Plasma Electrolytic Deposition of Al2O3 Antioxidation Coatings on Ti-45Al-8.5Nb Alloys. *Metals* **2018**, *8*, 308. [CrossRef]

Article

A Study on the Effect of Ambient Air Plasma Treatment on the Properties of Methylammonium Lead Halide Perovskite Films

Masoud Shekargoftar *, Jana Jurmanová and Tomáš Homola

R&D Centre for Low-Cost Plasma and Nanotechnology Surface Modifications (CEPLANT), Department of Physical Electronics, Faculty of Science, Masaryk University, Kotlářská 267/2, 611 37 Brno, Czech Republic

* Correspondence: mshekargoftar@mail.muni.cz; Tel.: +420-549-498-173

Received: 7 August 2019; Accepted: 5 September 2019; Published: 7 September 2019

Abstract: Organic-inorganic halide perovskite materials are considered excellent active layers in the fabrication of highly efficient and low-cost photovoltaic devices. This contribution demonstrates that rapid and low-temperature air-plasma treatment of mixed organic-inorganic halide perovskite film is a promising technique, controlling its opto-electrical surface properties by changing the ratio of organic-to-inorganic components. Plasma treatment of perovskite films was performed with high power-density ($25\ kW/m^2$ and $100\ W/cm^3$) diffuse coplanar surface barrier discharge (DCSBD) at 70 °C in ambient air. The results show that short plasma treatment time (1 s, 2 s, and 5 s) led to a relatively enlargement of grain size, however, longer plasma treatment time (10 s and 20 s) led to an etching of the surface. The band-gap energy of the perovskite films was related to the duration of plasma treatment; short periods (≤5 s) led to a widening of the band gap from ~1.66 to 1.73 eV, while longer exposure (>5 s) led to a narrowing of the band gap to approx. 1.63 eV and fast degradation of the film due to etching. Surface analysis demonstrated that the film became homogeneous, with highly oriented crystals, after short plasma treatment; however, prolonging the plasma treatment led to morphological disorders and partial etching of the surface. The plasma treatment approach presented herein addresses important challenges in current perovskite solar cells: tuning the optoelectronic properties and manufacturing homogeneous perovskite films.

Keywords: organic-inorganic halide perovskite; air plasma; plasma treatment; optoelectronic properties; morphology

---

1. Introduction

Organic-inorganic halide perovskites have recently attracted considerable attention, largely because of their outstanding opto-electronic qualities, among them a high absorption coefficient, tunable band-gap energy, and high charge-carrier mobility [1–4]. In recent years, the power conversion efficiency of perovskite solar cells (PSCs) has been improved from an initial 3% to over 25% through a variety of development technologies, including solvent engineering, chemical composition management, and the employment of different fabrication architecture within the cells [5,6]. One of the main challenges in the production of high-efficiency PSCs is the control of surface composition and morphology in such a way as to obtain perovskite films with smooth surfaces [7–9]. Organic-inorganic halide perovskite has been deposited and processed in many different ways, resulting in films of various surface compositions and morphology. For example, the average grain size of deposited perovskite films has been reported as varying from nanoscale to microscale [10]. The surface composition and morphology of a perovskite layer strongly influence its energy-level alignment, exciton diffusion length, mobility, lifetime, and generally affecting performance of the PSCs. Much effort has, therefore, been devoted to controlling the surface composition and morphology of perovskite films [11–14]. Recently, post-processing methods have been developed to improve their surface properties [15]. Thermal annealing of deposited perovskite films is one of the common processes that influence the quality of the film [16]. Surface passivation engineering has been demonstrated as having a remarkable effect on efficiency enhancement of the PSCs by the amelioration of defects and optimization of grain boundaries at the surface. In addition, surface properties of the other layers such as TiO_2 and PEDOT:PSS could significantly affect the performance of PCSs. For instance, Masood et al. studied the effect of TiO_2 thickness on the performance of PSCs [17]. Zhang et al. showed that treatment of the PEDOT:PSS buffer layer could enhance the efficiency of PSCs [18].

A wide range of procedures related to the post-treatment of perovskite films have been undertaken recently in the quest to produce high-quality perovskite and enhance the performance of PSCs. Microwave irradiation has proved a rapid and controllable post-treatment means of improving efficiency of the PSCs through the enhancement of crystallization [19]. Ren et al. reported on perovskite films annealed in an oxygen atmosphere for 12 h at various temperatures from room temperature to 85 °C [20]. Yang et al. reported that a time-temperature-dependent post-treatment (post-annealing) of a $CH_3NH_3PbI_xCl_{3-x}$ film led to a homogenous and highly oriented crystalline perovskite film that improved performance [21]. Plasma treatment has also shown great promise for surface treatment of perovskite films. Xiao et al. reported that low-pressure RF argon plasma treatment of the perovskite active layer can significantly improve the efficiency of PSCs [22]. The argon plasma treatment enhanced charge collection at the interface between the perovskite and electrode by altering the organic-to-inorganic ratio at the surface of perovskite layer. Plasma may also be employed for surface treatment of a wide range of materials. For example, UV/ozone plasma improves the optical and electrical characteristics of hybrid polymer-metal (PEN-Ag) electrodes in polymer solar cells [23]. CO_2 plasma treatment significantly heightens surface wettability, and thereby the electrical conductivity of TiO_2 transparent electrodes, improving the performance of the final device [24]. Ambient air plasma activation of the mesoscopic TiO_2 layer affects the efficiency of the PSC [25].

In the investigation presented herein, mixed organic-inorganic halide perovskite ($CH_3NH_3PbI_xCl_{3-x}$) films were deposited on quartz glass and then treated with low-temperature ambient-air surface plasma. The plasma was generated by diffuse coplanar surface dielectric barrier discharge (DCSBD), details of which appear in Černák et al. [26]. DCSBD ambient-air plasma has recently been tested for rapid and low-temperature calcination of TiO_2, Al_2O_3, and CeO_2 nanofibers [27–30], mineralization of TiO_2 mesoporous flexible photoanodes [31,32], and arrangement within an enclosed system with hydrogen gas to allow swift production of flexible reduced graphene oxide electrodes on PET foils [33]. Here, $CH_3NH_3PbI_xCl_{3-x}$ perovskite films were spin-coated on to quartz glass and the effects of plasma treatment on the morphological and optoelectronic properties of these films were investigated. The results showed that, plasma treatment effectively removed organic methylammonium cations (CH_3NH_3 or MA) from the surface, however prolonging the plasma treatment lead to partially etching the surface. The work demonstrates that optoelectronic qualities of perovskite films, as measured by UV–VIS and ultraviolet photoelectron spectroscopy, can respond significantly to the surface composition and morphology. The chemical composition of the mixed halide perovskite films was evaluated by X-ray photoelectron spectroscopy (XPS) and Fourier transform infrared spectroscopy attenuated total reflectance (ATR-FTIR). Short plasma treatment (≤5 s) led to homogeneous perovskite films of larger grain size, as detected by X-ray diffraction (XRD) and scanning electron microscopy (SEM).

2. Experimental Section

2.1. Film Deposition

Perovskite precursor ink (a mixture of methylammonium iodide (MAI) and lead chloride (PbCl) dissolved in dimethyl formamide) was purchased from Ossila, UK. Prior to deposition, the perovskite solution was heated to 70 °C for 2 h. A quartz glass sheet was cut into small pieces (1 cm × 1 cm), then ultrasonically cleaned for 6 min in acetone, followed by isopropyl alcohol and deionized water. The samples were pre-treated with air plasma for 16 s (the optimum pre-treatment time for obtaining a uniform film on the substrate). After plasma pre-treatment, the samples were placed in a spin coater and perovskite ink was deposited on the substrate at a speed of 3000 rpm; the volume of the perovskite ink on each sample was 10 μL. All samples were then placed on a hotplate set at a temperature of 80 °C for 2 h. All the processes of the sample and ink deposition are performed in ambient-air conditions ranging from 20 °C to 24 °C and at a relative humidity of ≤35%.

2.2. Plasma Processing

Ambient-air plasma was generated by dielectric barrier discharge with a coplanar configuration of the electrodes using a diffuse coplanar surface barrier discharge (DCSBD) unit manufactured by Roplass s.r.o. (Brno, Czech Republic). The power density of plasma was maintained at 25 kW/m^2 and 100 W/cm^3. The deposited films were exposed to plasma for various treatment times ranging from 1 s to 20 s. Further, the deposited samples were treated at a range of distances between sample and plasma (from 0.1 to 0.7 mm). A distance of 0.6 mm was selected for the experiment because this was considered the optimal distance in light of the results (Figure S1, supplementary materials).

2.3. Characterization

A PerkinElmer Lambda 1050 UV–VIS spectrometer (Akron, OH, USA) was employed to measure the optical properties of the perovskite films at wavelengths from 380 nm to 800 nm. X-ray diffraction studies (XRD) were conducted using a Rigaku Smartlab (The Woodlands, TX, USA), with CuKα (λ = 1.54059A°) radiation. Ultraviolet photoelectron spectroscopy (UPS) was performed by Kratos Supra (Wharfside, Manchester, UK) under normal emission using He I (21.22 eV). The spectra were calibrated at the Fermi edge, which is the energy of the fastest emitted electron. To separate the secondary electron (SE) onset of the samples from that of the detector, a negative bias was applied to the samples. Chemical analysis of the samples was carried out by X-ray photoelectron spectroscopy (XPS) using an ESCALAB 250Xi of Al Kα X-ray source (Thermo Fisher Scientific, Loughborough, UK). The XPS spectra were acquired from two spots (650 μm) with a take-off angle of 90° in 10^{-8} mbar vacuum at 20 °C. Additional chemical information from the samples was obtained by using attenuated total reflectance Fourier transform infrared (ATR-FTIR) with a Bruker Vertex 80 V spectrometer (Brno, Czech Republic). The perovskite films were measured in attenuation total reflection (ATR) mode using a diamond crystal at a pressure of 2.51 hPa, spectral range from 4000 to 600 cm^{-1} and a resolution of 4 cm^{-1}. A MIRA3 scanning electron microscope (Tescan, Czech Republic) was used to evaluate the morphology of the perovskite films before and after plasma treatment. To minimize the influence of environment on the plasma-treated perovskite films, the time between plasma treatment and surface characterization was always less than 1 h and samples were transported under nitrogen atmosphere with neither oxygen nor moisture.

3. Results and Discussion

XRD patterns for perovskite films before and after plasma treatment appear in Figure 1a. The peak located at 14.1° corresponding to the (110) plane belongs to the MAPbI$_{3-x}$Cl$_x$ phase and the intensity of this peak decreased after plasma treatment but each treatment time led to different intensity. A notable diffraction peak at 15.5° corresponding to the (001) plane of the perovskite film appeared after a short plasma treatment time (1 s), indicating the formation of a MAPbCl$_3$ perovskite structure. Prolonging the plasma treatment time led to increases in the intensity of this peak. Another peak of MAPbCl$_3$ located at 16.5° corresponding to the (012) plane appeared after the plasma treatment. The peaks located at 28.5°, 29.3°, and 31.8° correspond to the (220), (002), and (310) planes of MAPbI$_3$, respectively. The peak of (310) shifted to a lower angle after plasma treatment, a change perhaps related to an expanded crystal lattice. The intensity of the peak at 28.5° increased after plasma treatment which could be related to the enhancement of crystallinity within the perovskite film [34].

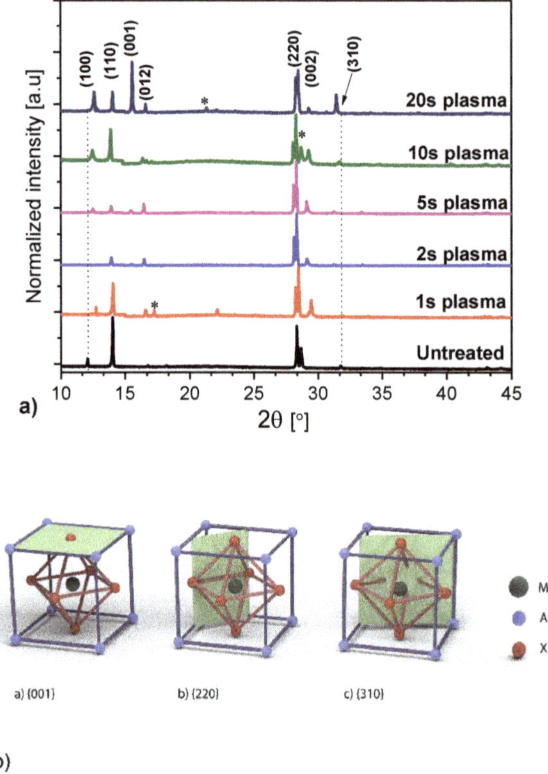

Figure 1. (**a**) XRD pattern of perovskite film before and after plasma treatment. Denoted peaks with * correspond to the intermediate phase. (**b**) The location of some perovskite planes in the unit cell.

A slight diffraction of the PbI_2 peaks at the (100) plane appeared in the XRD pattern. Figure 1b shows the location of some perovskite planes in the unit cell. The presence of the PbI_2 peaks on the "untreated" sample refers to the incomplete conversion to the perovskite phase, and the peak shifted to a lower angle after plasma treatment. The PbI_2 peaks detected on the plasma-treated films could be attributed to partial decomposition of the film. According to the Dubey et al. [35], the decomposition of perovskite film originates from moisture in the air. In addition, PbI_2 peak showed different trends after plasma treatment: short plasma treatment ($\leq$5 s) led to a decrease of the PbI_2 intensity and prolonging the plasma treatment increased the intensity of PbI_2. Furthermore, the calculated sizes of the PbI_2 crystallite using the Scherrer equation (Figure S2) showed the same trend of the PbI_2 peak before and after plasma treatment.

Figure 2 shows the UV–VIS absorption spectra of the perovskite film before and after plasma treatment. These demonstrate that plasma treatment led to shifts in the spectra of the perovskite film, varying with treatment times. A blue-shift of the absorption band edge occurred after short plasma treatment ($\leq$5 s). The absorption band edge of the "untreated" perovskite film, the starting location at approx. 743 nm, shifted to approx. 713–716 nm after short plasma treatments (1 s and 2 s). The band edge after 5 s appeared at approx. 718 nm. Prolonging the treatment moved the band edge towards surprisingly higher wavelengths (red-shift), whereas 10 s of plasma treatment shifted the absorption band edge to approx. 757 nm. After 20 s of treatment, the band edge lay at a wavelength of 760 nm.

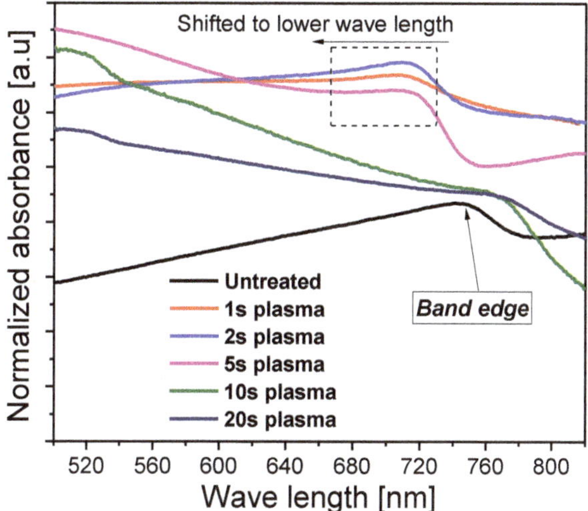

Figure 2. UV–VIS absorbance spectra of the mixed-halide perovskite film before and after plasma treatment.

Secondary onset (SE onset) and valance bands of the perovskite films were measured by UPS; the results appear in Figure 3 and Figure S3, respectively.

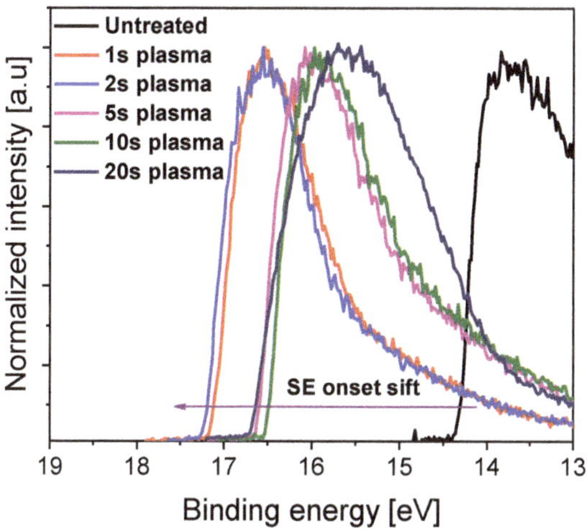

Figure 3. Secondary electron (SE) onset of the mixed-halide perovskite film obtained by UPS before and after plasma treatment.

The SE onset and valance band maximum (VBM) of the "untreated" perovskite films on quartz glass were located at 14.36 eV and 4.24 eV, respectively. The SE onset of the perovskite film shifted towards higher binding energy after plasma treatment and the maximum shift appeared after treatment for 2 s, with the SE onset located at a binding energy of 17.27 eV. Measurement of UPS indicated that the VBM of the perovskite films reacted differently to shorter and longer treatment times. Short plasma treatment shifted the valence bands to lower binding energy, with the VBM located at 2.39–2.46 eV after 1–5 s. In contrast, prolonged plasma treatment resulted in the VBM at higher binding energy, with the VBM located at 6.63 and 10.41 eV after plasma treatment for 10 s and 20 s, respectively.

The conduction band edge, or minimum conduction band (mCB), of the perovskite film was estimated by means of band gap energy and maximum valance band. The band gap energy (E_g) can be calculated by using hc/λ, where 'h' is Plank's constant, c is the speed of the light in vacuum, and λ is the wavelength corresponding to the band edge within the UV–VIS absorption spectra. Figure 4a shows the variation of the energy bands (including E_g, VBM, and mCB) of the perovskite films during plasma treatment, together with corresponding photographs taken after deposition. In general, the band gap shifts relate to changes in carrier concentration [36], meaning that a band gap shift toward larger values corresponds to a higher carrier concentration. The band gap widening after short plasma treatment ($\leq$5 s) could, thus, be attributed to a higher concentration of charge carriers within the perovskite film. Further, the higher energy band was associated with decreases in VBM and mCB to lower energy levels. However, longer plasma treatment led to narrowing of the band gap of the mixed halide perovskite film, associated with higher energy levels of VBM and mCB.

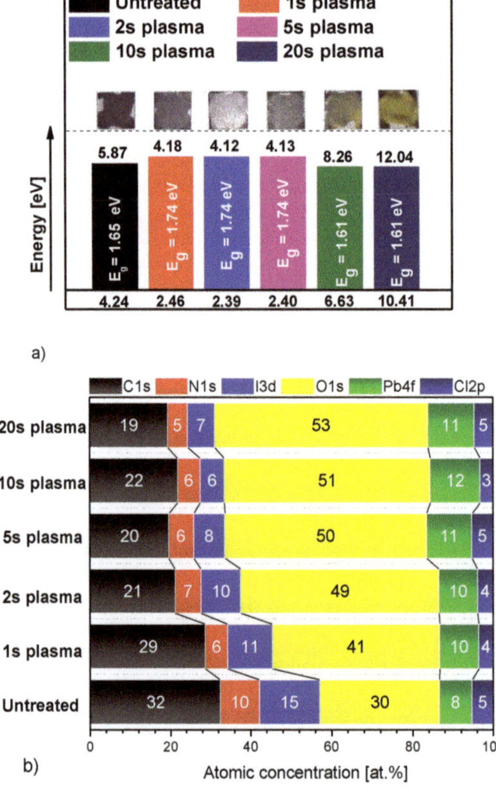

Figure 4. (**a**) Band energy diagram, with corresponding views from above of the mixed-halide perovskite film before and after plasma treatment. (**b**) Atomic concentration of elements in perovskite films calculated from XPS survey spectra.

It is well known that ambient air plasma treatment removes organic contamination from surfaces and induce oxygen-containing functional groups to the surface [37]. It was, therefore, anticipated that plasma treatment would lead to a decrease in the organic part of the mixed-halide perovskite film, e.g., in methylammonium cations ($CH_3NH_3^+$). Evaluation of the chemical stoichiometry of the perovskite film before and after plasma treatment was performed by means of XPS measurement. Figure 4b summarizes the XPS results, including atomic concentration as a function of the plasma treatment time. The atomic concentration of the C1s peak decreased from approx. 32.4 at%, corresponding to the "untreated" perovskite film, to 28.7 at% after short plasma treatment (1 s). Plasma treatment for 2 s led to a concentration of the C1s peak to 21.2 at%, while prolonging the plasma treatment to 20 s led to a concentration of C1s of 19.3 at%. Longer plasma treatment times >5 s had no further effect on the composition of the perovskite surface probably due to homogeneous etching of the perovskite film, which is a different process than etching of the outermost layer of the surface. The XPS results disclosed that the lower concentration of C1s is accompanied by a dramatic decrease in the N1s and I3d concentrations after plasma treatment. In addition, the atomic concentration of oxygen was enhanced. High energy species of plasma could etch the surface of the perovskite film which broke the chemical bonds of the organic and halide components. The bonding of Pb is too strong for plasma to break it but contamination removal from the outermost layer of surface led to an increase in Pb concentration. The concentration of Pb4f increased from 8.1 to 9.7 at% after short plasma treatment (1 s). Prolonging the plasma treatment led to an increase in the Pb4f concentration. The rise in lead (Pb4f) concentration was related to the removal of organic contamination from the surface. Lead concentration remained constant after plasma treatment for 5 s, indicating that plasma penetration to the bulk of the mixed-halide perovskite films was negligible.

XPS revealed that ambient air plasma treatment partially removed carbon contamination, methylammonium cations, and halide anions from the surface of the mixed-halide perovskite films. ATR-FTIR measurement was employed further to investigate the effects of plasma treatment on the chemistry of mixed-halide perovskite; the ATR-FTIR patterns of the perovskite films appear in Figure 5. The peaks located at 1200–1700 cm^{-1} are related to C–N and C=O stretch, and the intensity of these peaks decreased dramatically after plasma treatment. In addition, the peak located at 2700–3000 cm^{-1} corresponded to C–H stretching, and its intensity fell to around zero after plasma treatment for 1 s. The treatment also reduced the intensity of the peak located at 3000–3300 cm^{-1}, corresponding to N–H/O–H stretching. These results correlated well with the atomic concentrations obtained by XPS, showing that plasma treatment led to effective removal of the organic part of the mixed-halide perovskite films. Chemical analysis of the perovskite films indicated that ambient-air plasma treatment cleans surfaces and forms lead-rich perovskite films.

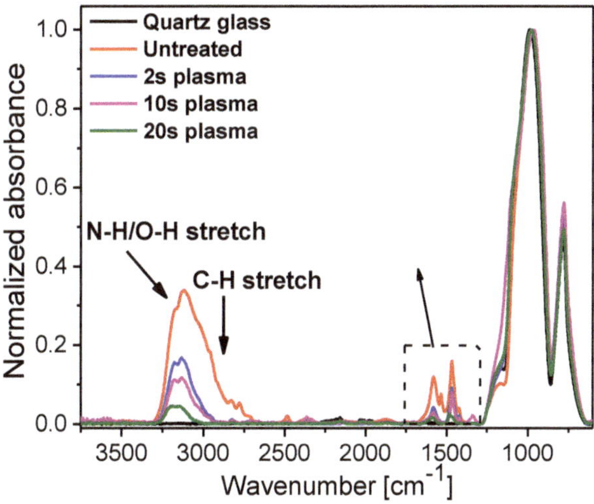

Figure 5. ATR-FTIR pattern of mixed-halide perovskite film before and after plasma treatment.

Scanning electron microscopy (SEM) was employed to evaluate the effect of plasma treatment on the surface morphology of the perovskite films (Figure 6). The surface of the "untreated" mixed-halide perovskite film shows the inhomogeneity and compactness of the grains. The non-homogeneous "untreated" surface may be explained by evaporation of the solvent of the perovskite precursor from the substrate. The perovskite precursor ink was a mixture of methylammonium iodide and lead chloride dissolved in dimethyl formamide. Once the perovskite had been spin-coated onto the substrate, the solvent evaporated very rapidly and generated non-uniform perovskite crystals [38]. Short plasma treatment (1 s) led to higher homogeneity of the surface as grain boundaries appeared. However, the sample treated for 1 s exhibited a number of shrunken grain boundaries. The same phenomena were reported by Cao et al. [19], who noted that defects in the grain boundaries of perovskite films appeared after microwave irradiation treatment, related to the decomposition of the perovskite film. These results accord with the XRD analysis, which revealed partial decomposition of the mixed-halide perovskite film after short plasma treatment of 1 s. The cracks became less after plasma treatment for 2 s and 5 s. Prolonging plasma treatment time after 5 s led to pin-holing and separation of the grains at the surface. The calculated average grain size from the SEM results are summarized in Table 1. It appears that short plasma treatment (≤5 s) led to extended grain sizes of the mixed-halide perovskite; the average grain size corresponding to the "untreated" film is ~0.92 and increased to 1.93 nm after 5 s plasma treatment. However, longer plasma treatments of 10 s and 20 s decreased the average grain size to ~0.78 nm and 0.54 nm, respectively, which can be related to the etching of the surface after longer plasma treatment. The thickness of the of mixed-halide perovskite films before and after plasma treatment was also investigated (Figure S4). The results demonstrated that, after 1 s and 2 s of exposure time, the thickness of the films decreased from approx. 229 nm to 225 nm and 218 nm. This may be related to the etching of organic contamination from the surface, and even of the organic component of the perovskite film itself. Longer plasma treatments (>5 s) etched more of the surface and thickness decreased to approx. 200 nm after plasma treatment for 20 s, suggesting that the upper layer of the surface was removed.

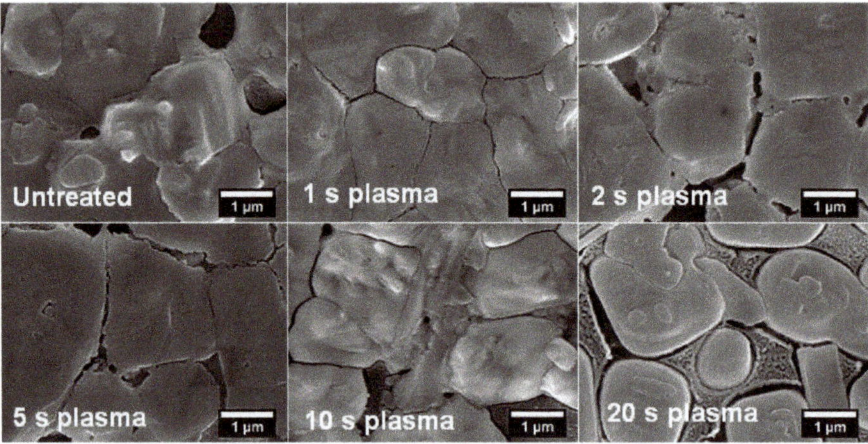

Figure 6. SEM images of mixed-halide perovskite films before and after plasma treatment.

Table 1. Average grain size of the mixed-halide perovskite films before and after plasma treatment.

Sample	Grain Size (µm)
Untreated	0.92
1 s plasma	1.36
2 s plasma	1.81
5 s plasma	1.93
10 s plasma	0.78
20 s plasma	0.54

4. Conclusions

Ambient air low-temperature plasma treatment was successfully applied to perovskite film for the first time. The effects of such plasma treatment on the opto-electronic properties, crystallization, and morphology of the perovskite film were investigated. Ambient-air plasma led to remarkable changes in the energy bands of the mixed-halide perovskite in a manner dependent on treatment time, i.e., short plasma treatment ($\leq$5 s) led to widening the band gap and longer plasma treatment (>5 s) led to narrowing the band gap. Plasma treatment can control the morphology of the perovskite film by changing grain boundaries and crystalline structure, leading to a uniform surface after short plasma treatment. However, prolonging plasma treatment leads to the separation of the grains and the film becomes non-homogeneous. The differences between the results of short and long plasma treatment were related to the etching of the perovskite film. Short plasma treatment had a minor etching effect on the surface, in contrast to longer treatment that etched the surface and penetrated to the bulk of the perovskite film and consequently led to slight etching of the surface. This contribution demonstrates that plasma treatment using DCSBD is a promising post-treatment technique for control of the surface properties of mixed-halide perovskite films. In addition, using ambient air as an operational gas, together with short treatment times, render plasma treatment by DCSBD a low-cost and rapid means of processing perovskite films.

Supplementary Materials: The following are available online at http://www.mdpi.com/2075-4701/9/9/991/s1, Figure S1: Photo of plasma treated perovskite films with different distances, Figure S2: Crystallite size of PbI_2 at plate of (100) before and after plasma treatment, Figure S3: Variation of valance bands of mixed-halide perovskite film during plasma treatment, Figure S4: Variation the thickness of mixed-halide perovskite during plasma treatment time.

Author Contributions: Experiments and data analysis and draft preparation, M.S.; Data analysis and writing-editing, M.S. and T.H.; SEM measurements and morphology analysis, J.J., M.S.

Funding: This research was supported by project LO1411 (NPU I) funded by Ministry of Education Youth and Sports of Czech Republic. Part of the work was carried out at CEITEC Nano Research Infrastructure (MEYS CR, 2016–2019). The authors would like to thank the Grant Agency of Masaryk University. The work of Masoud Shekargoftar was also supported by Brno Ph.D. Talent Scholarship, funded by the City of Brno.

Acknowledgments: The authors would like to thank David Pavliňák, Monika Stupavská, Pavel Franta and Ondrej Caha for the acquired ATR-FTIR, XPS and UV-Vis data and XRD data. Tony Long (Svinošice) helped work up the English.

Conflicts of Interest: The authors declare no conflict of interest.

References

1. Ava, T.T.; Al Mamun, A.; Marsillac, S.; Namkoong, G. A Review: Thermal Stability of Methylammonium Lead Halide Based Perovskite Solar Cells. *Appl. Sci.* **2019**, *9*, 188. [CrossRef]
2. Li, H.; Zhu, K.; Zhang, K.; Huang, P.; Li, D.; Yuan, L.; Cao, T.; Sun, Z.; Li, Z.; Chen, Q.; et al. 3,4-Dihydroxybenzhydrazide as an additive to improve the morphology of perovskite films for efficient and stable perovskite solar cells. *Org. Electron.* **2019**, *66*, 47–52. [CrossRef]
3. Wang, R.; Mujahid, M.; Duan, Y.; Wang, Z.-K.; Xue, J.; Yang, Y. A Review of Perovskites Solar Cell Stability. *Adv. Funct. Mater.* **2019**, *1808843*, 1–25. [CrossRef]
4. Singh, R.; Singh, P.K.; Bhattacharya, B.; Rhee, H.-W. Review of current progress in inorganic hole-transport materials for perovskite solar cells. *Appl. Mater. Today* **2019**, *14*, 175–200. [CrossRef]
5. Huang, F.; Li, M.; Siffalovic, P. Environmental Science From scalable solution fabrication of perovskite films towards commercialization of solar cells. *Energy Environ. Sci.* **2019**, *12*, 518–549. [CrossRef]
6. Adjokatse, S.; Kardula, J.; Fang, H.-H.; Shao, S.; Brink, G.H.T.; Loi, M.A. Effect of the Device Architecture on the Performance of $FA_{0.85}MA_{0.15}PbBr_{0.45}I_{2.55}$ Planar Perovskite Solar Cells. *Adv. Mater. Interfaces* **2019**, *6*, 1801667. [CrossRef]
7. Correa-Baena, J.-P.; Saliba, M.; Buonassisi, T.; Grätzel, M.; Abate, A.; Tress, W.; Hagfeldt, A. Promises and challenges of perovskite solar cells. *Science* **2017**, *358*, 739–744. [CrossRef]

8. Zhou, Y.; Yang, M.; Game, O.S.; Wu, W.; Kwun, J.; Strauss, M.A.; Yan, Y.; Huang, J.; Zhu, K.; Padture, N.P. Manipulating Crystallization of Organolead Mixed-Halide Thin Films in Antisolvent Baths for Wide-Bandgap Perovskite Solar Cells. *ACS Appl. Mater. Interfaces* **2016**, *8*, 2232–2237. [CrossRef]

9. Uribe, J.I.; Ciro, J.; Montoya, J.F.; Osorio, J.; Jaramillo, F. Enhancement of Morphological and Optoelectronic Properties of Perovskite Films by CH_3NH_3Cl Treatment for Efficient Solar Minimodules. *ACS Appl. Energy Mater.* **2018**, *1*, 1047–1052. [CrossRef]

10. Chu, Z.; Yang, M.; Schulz, P.; Wu, D.; Ma, X.; Seifert, E.; Sun, L.; Li, X.; Zhu, K.; Lai, K. Impact of grain boundaries on efficiency and stability of organic-inorganic trihalide perovskites. *Nat. Commun.* **2017**, *8*, 2230. [CrossRef]

11. Yu, Y.; Yang, S.; Lei, L.; Cao, Q.; Shao, J.; Zhang, S.; Liu, Y. Ultrasmooth Perovskite Film via Mixed Anti-Solvent Strategy with Improved Efficiency. *ACS Appl. Mater. Interfaces* **2017**, *9*, 3667–3676. [CrossRef] [PubMed]

12. Huang, L.-B.; Su, P.-Y.; Liu, J.-M.; Huang, J.-F.; Chen, Y.-F.; Qin, S.; Guo, J.; Xu, Y.-W.; Su, C.-Y. Interface engineering of perovskite solar cells with multifunctional polymer interlayer toward improved performance and stability. *J. Power Sources* **2018**, *378*, 483–490. [CrossRef]

13. Eperon, G.E.; Burlakov, V.M.; Docampo, P.; Goriely, A.; Snaith, H.J. Morphological control for high performance, solution-processed planar heterojunction perovskite solar cells. *Adv. Funct. Mater.* **2014**, *24*, 151–157. [CrossRef]

14. Shi, D.; Adinolfi, V.; Comin, R.; Yuan, M.; Alarousu, E.; Buin, A.; Chen, Y.; Hoogland, S.; Rothenberger, A.; Katsiev, K.; et al. Low trap-state density and long carrier diffusion in organolead trihalide perovskite single crystals. *Science* **2015**, *347*, 519–522. [CrossRef] [PubMed]

15. Yoo, H.-S.; Park, N.-G. Post-treatment of perovskite film with phenylalkylammonium iodide for hysteresis-less perovskite solar cells. *Sol. Energy Mater. Sol. Cells* **2018**, *179*, 57–65. [CrossRef]

16. Dong, G.; Xia, D.; Yang, Y.; Sheng, L.; Su, T.; Fan, R.; Shi, Y.; Wang, J. Inverted thermal annealing of perovskite films: A method for enhancing photovoltaic device efficiency. *RSC Adv.* **2016**, *6*, 44034–44040. [CrossRef]

17. Masood, M.T.; Weinberger, C.; Sarfraz, J.; Rosqvist, E.; Sandén, S.; Sandberg, O.J.; Vivo, P.; Hashmi, G.; Lund, P.D.; Österbacka, R.; et al. Impact of Film Thickness of Ultrathin Dip-Coated Compact TiO_2 Layers on the Performance of Mesoscopic Perovskite Solar Cells. *ACS Appl. Mater. Interfaces* **2017**, *9*, 17906–17913. [CrossRef] [PubMed]

18. Zhang, R.; Ling, H.; Lu, X.; Xia, J. The facile modification of PEDOT: PSS buffer layer by polyethyleneglycol and their effects on inverted perovskite solar cell. *Sol. Energy* **2019**, *186*, 398–403. [CrossRef]

19. Cao, Q.; Yang, S.; Gao, Q.; Lei, L.; Yu, Y.; Shao, J.; Liu, Y. Fast and Controllable Crystallization of Perovskite Films by Microwave Irradiation Process. *ACS Appl. Mater. Interfaces* **2016**, *8*, 7854–7861. [CrossRef]

20. Ren, Z.; Ng, A.; Shen, Q.; Gokkaya, H.C.; Wang, J.; Yang, L.; Yiu, W.-K.; Bai, G.; Djurišić, A.B.; Leung, W.W.-F.; et al. Thermal Assisted Oxygen Annealing for High Efficiency Planar CH3NH3PbI3 Perovskite Solar Cells. *Sci. Rep.* **2014**, *4*, 6752. [CrossRef]

21. Yang, Y.; Feng, S.; Li, M.; Xu, W.; Yin, G.; Wang, Z.; Sun, B.; Gao, X. Annealing Induced Re-crystallization in $CH_3NH_3PbI_{3-x}Cl_x$ for High Performance Perovskite Solar Cells. *Sci. Rep.* **2017**, *7*, 46724. [CrossRef] [PubMed]

22. Xiao, X.; Bao, C.; Fang, Y.; Dai, J.; Ecker, B.R.; Wang, C.; Lin, Y.; Tang, S.; Liu, Y.; Deng, Y.; et al. Argon Plasma Treatment to Tune Perovskite Surface Composition for High Efficiency Solar Cells and Fast Photodetectors. *Adv. Mater.* **2018**, *30*, 1705176. [CrossRef] [PubMed]

23. Zheng, W.; Lin, Y.; Zhang, Y.; Yang, J.; Peng, Z.; Liu, A.; Zhang, F.; Hou, L. Dual Function of UV/Ozone Plasma-Treated Polymer in Polymer/Metal Hybrid Electrodes and Semitransparent Polymer Solar Cells. *ACS Appl. Mater. Interfaces* **2017**, *9*, 44656–44666. [CrossRef] [PubMed]

24. Wang, K.; Zhao, W.; Liu, J.; Niu, J.; Liu, Y.; Ren, X.; Feng, J.; Liu, Z.; Sun, J.; Wang, D.; et al. CO_2 Plasma-Treated TiO_2 Film as an Effective Electron Transport Layer for High-Performance Planar Perovskite Solar Cells. *ACS Appl. Mater. Interfaces* **2017**, *9*, 33989–33996. [CrossRef] [PubMed]

25. Masood, M.T.; Weinberger, C.; Qudsia, S.; Rosqvist, E.; Sandberg, O.J.; Nyman, M.; Sandén, S.; Vivo, P.; Aitola, K.; Lund, P.D.; et al. Influence of titanium dioxide surface activation on the performance of mesoscopic perovskite solar cells. *Thin Solid Films* **2019**, *686*, 137418. [CrossRef]

26. Černák, M.; Kováčik, D.; Ráhel', J.; St'ahel, P.; Zahoranová, A.; Kubincová, J.; Tóth, A.; Černáková, L. Generation of a high-density highly non-equilibrium air plasma for high-speed large-area flat surface processing. *Plasma Phys. Control. Fusion* **2011**, *53*, 124031. [CrossRef]

27. Medvecká, V.; Kováčik, D.; Zahoranová, A.; Černák, M. Atmospheric pressure plasma assisted calcination by the preparation of TiO$_2$ fibers in submicron scale. *Appl. Surf. Sci.* **2018**, *428*, 609–615. [CrossRef]
28. Mudra, E.; Streckova, M.; Pavlinak, D.; Medvecka, V.; Kovacik, D.; Kovalcikova, A.; Zubko, P.; Girman, V.; Dankova, Z.; Koval, V.; et al. Development of Al$_2$O$_3$ electrospun fibers prepared by conventional sintering method or plasma assisted surface calcination. *Appl. Surf. Sci.* **2017**, *415*, 90–98. [CrossRef]
29. Kováčik, D.; Medvecká, V.; Tucekova, Z.; Zahoranova, A.; Cernak, M. Atmospheric pressure plasma assisted calcination of composite submicron fibers. *Eur. Phys. J. Appl. Phys.* **2016**, *75*, 24715.
30. Medvecká, V.; Kováčik, D.; Zahoranová, A.; Stupavská, M.; Černák, M. Atmospheric pressure plasma assisted calcination of organometallic fibers. *Mater. Lett.* **2016**, *162*, 79–82. [CrossRef]
31. Shekargoftar, M.; Dzik, P.; Stupavska, M.; Pavlinak, D.; Homola, T. Mineralization of flexible mesoporous TiO$_2$ photoanodes using two low-temperature DBDs in ambient air. *Contrib. Plasma Phys.* **2018**, *59*, 102–110. [CrossRef]
32. Homola, T.; Shekargoftar, M.; Dzik, P.; Krumpolec, R.; Ďurašová, Z.; Veselý, M.; Černák, M. Low-temperature (70 °C) ambient air plasma-fabrication of inkjet-printed mesoporous TiO$_2$ flexible photoanodes. *Flex. Print. Electron.* **2017**, *2*, 35010. [CrossRef]
33. Homola, T.; Pospišil, J.; Krumpolec, R.; Souček, P.; Dzik, P.; Weiter, M.; Černák, M. Atmospheric Dry Hydrogen Plasma Reduction of Inkjet-Printed Flexible Graphene Oxide Electrodes. *ChemSusChem* **2018**, *11*, 941–947. [CrossRef] [PubMed]
34. Zhou, L.; Chang, J.; Liu, Z.; Sun, X.; Lin, Z.; Chen, D.; Zhang, C.; Zhang, J.; Hao, Y. Enhanced planar perovskite solar cell efficiency and stability using a perovskite/PCBM heterojunction formed in one step. *Nanoscale* **2018**, *10*, 3053–3059. [CrossRef] [PubMed]
35. Dubey, A.; Adhikari, N.; Mabrouk, S.; Wu, F.; Chen, K.; Yang, S.; Qiao, Q. A strategic review on processing routes towards highly efficient perovskite solar cells. *J. Mater. Chem. A* **2018**, *6*, 2406–2431. [CrossRef]
36. Lu, J.G.; Fujita, S.; Kawaharamura, T.; Nishinaka, H.; Kamada, Y.; Ohshima, T.; Ye, Z.Z.; Zeng, Y.J.; Zhang, Y.Z.; Zhu, L.P.; et al. Carrier concentration dependence of band gap shift in *n*-type ZnO:Al films. *J. Appl. Phys.* **2007**, *101*, 083705. [CrossRef]
37. Shekargoftar, M.; Krumpolec, R.; Homola, T. Enhancement of electrical properties of flexible ITO/PET by atmospheric pressure roll-to-roll plasma. *Mater. Sci. Semicond. Process.* **2018**, *75*, 95–102. [CrossRef]
38. Li, X.; Li, L.; Ma, Z.; Huang, J.; Ren, F. Low-cost synthesis, fluorescent properties, growth mechanism and structure of CH$_3$NH$_3$PbI$_3$ with millimeter grains. *Optik* **2017**, *142*, 293–300. [CrossRef]

Article

Spatial Distribution of Wettability in Aluminum Surfaces Treated with an Atmospheric-Pressure Remote-Plasma

José Muñoz *, Rocío Rincón and María Dolores Calzada

Laboratorio de Innovación en Plasmas, Edificio Einstein (C2), Campus de Rabanales, Universidad de Córdoba, 14071 Córdoba, Spain
* Correspondence: jmespadero@uco.es; Tel.: +34-957-218-627

Received: 14 August 2019; Accepted: 26 August 2019; Published: 27 August 2019

Abstract: The use of atmospheric-pressure remote plasmas (postdischarge) sustained in argon and argon–nitrogen for the treatment of aluminum surfaces has been studied to better understand the underlying mechanisms responsible for cleaning and activating the surfaces. The effect of the gas composition, treatment distance, and speed on the hydrophilicity of commercial aluminum samples has been studied using the sessile drop method to build spatial profiles of the treated zones. In the case of argon–nitrogen postdischarges, neither the distance to the plasma end ($2 < z < 6$ cm) nor the treatment speed ($2500 < v < 7500$ μm/s) had a significant impact in the spot radius of the treatment, remaining approximately constant around 6–7 mm. This result seems to indicate that the postdischarge experiments a little expansion at the exit of the tube in which the discharge was created but its action can be considered highly-directional. This fact is essential for the possible industrial implementation of the procedure described in this research. These results have been analyzed together with the composition of active species in the postdischarge by using optical emission spectroscopy, revealing that long lived nitrogen species are required to significantly increase the wettability of the aluminum surfaces.

Keywords: aluminum; surface; plasma; nitrogen; postdischarge; atmospheric pressure; wettability

---

1. Introduction

Among the different materials used in the industry, metals and their alloys stand out due to their outstanding properties such as thermal and electric conductivity, malleability or ductility which make them of special interest to be utilized in sectors like electronics, food packaging, automotive, aeronautics, or construction. Aluminum is one of the most used metals, and from the environmental point of view, it is 100% recyclable without the loss of any of its properties, thus guaranteeing the development of sustainable technologies on the long term. However, this strategic material is often exposed to harsh environments that provoke fast oxidation and/or corrosion of its surface. For example, in the development of lithium-ion batteries, it is common to use aluminum for the manufacture of the current collector, which is exposed to the pitting corrosion of the electrolytes of the battery, especially during long work cycles, like those required for their use in electric and/or hybrid cars. This deterioration can cause an increase in the impedance and self-discharge of the battery, as well as loss of its energy storage capability [1]. Similarly, aluminum surfaces can be easily damaged by corrosion due to chlorine ions, which is of great importance during the continue exposure to environmental agents of aluminum components in construction, naval, and aerospace industries [2].

Therefore, the development of a correct protection on aluminum surfaces, e.g., by depositing a protective film on the metallic material, becomes crucial to the achievement of significant advances in the demanding construction industry or the implementation of energy storage. Nevertheless, the inclusion of a protective film on aluminum surfaces requires an effective adhesion between them. Among other methodologies, plasma technology has demonstrated its capability to improve the adherence of different material surfaces by inducing changes on their surface energy and thus on their wettability behavior [3–5]. The role of the plasma in this process can be understood as an initial conditioning of the surface by two different approaches. On the one hand, energetic plasma species can remove impurities from the metal surface thus cleaning the surface and, on the other hand, during the plasma process, new surface functional groups, e.g., hydroxyl groups, can be inserted or deposited, thus activating the surface.

Dielectric Barrier Discharges (DBDs) have been used for the activation of surfaces of steel [6], chromium [7], and aluminum [8–10] because their geometry is more suitable for treating flat surfaces in continuous mode. However, some studies have explored the possibility of using other plasma sources for this purpose, such as radiofrequency or microwave plasmas [11–14], more suitable for the treatment of small and/or irregular surfaces.

In a previous investigation [14], the remote plasma (postdischarge) of argon and argon–nitrogen microwave plasmas was used for cleaning and activating the surface of commercial aluminum samples. The action of the postdischarge on the considered surfaces provoked a significant increase of the hydrophilicity and surface energy from 37 to values ranging from 69 (Ar postdischarge) to 77 mJ/m^2 (Ar-N$_2$ postdischarge)—values comparable and even larger than those required by the industrial processing standards for the deposition and adhesion of films on the metal surfaces. Besides, the effectiveness of the postdicharge treatment for distances up to 5 cm was tested; this distance being larger than those typically allowed by Dielectric Barrier Discharge (DBD) by an order of magnitude. In addition, the use of microwave plasmas allows for treating non-planar geometries, an advantage with respect to DBD discharges, which are limited to the treatment of flat surfaces.

The present work is a continuation of our previous investigation on the action of the microwave-plasma postdischarge to clean and activate aluminum metal surfaces [14]. The current research is devoted to analyzing the spatial resolution of the treatment carried out by the Ar-N$_2$ postdischarge as well as the surface mechanism during the postdischarge action on the aluminum surface.

2. Experimental and Methods

Figure 1a shows a diagram of the experimental setup used for creating and treating aluminum samples; a picture of the plasma and postdischarge can be observed in Figure 1b.

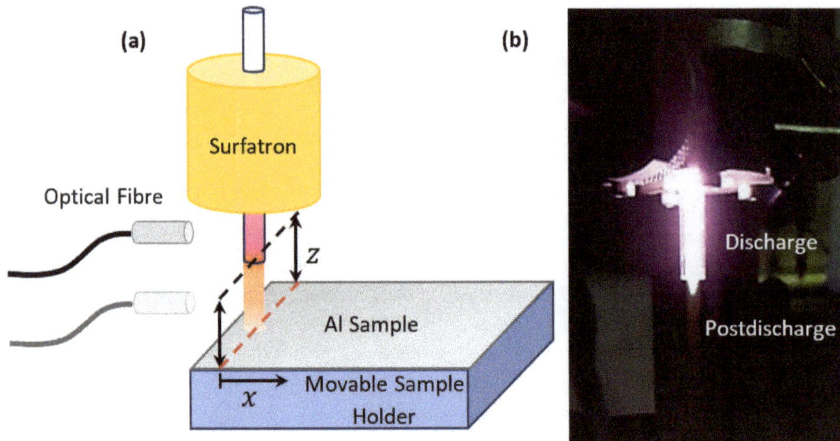

Figure 1. Experiment setup (**a**) and detail of an Ar-N$_2$ discharge and (**b**) postdischarge.

A Sairem GMP 03 k/SM microwave (2.45 GHz) generator (Sairem Ibérica S.L., Barcelona, Spain) supplied power of 150 W in continuous mode to create and maintain the surface-wave plasmas and a surfatron [15] was used as energy coupling device. Surface-wave discharges were created in quartz tubes of 2 and 3 mm of inner and outer radius, respectively, opened to the atmosphere at one of its ends. This tube extends 3 cm from the launching gap of the surfatron and the postdischarge expanded out the tube (Figure 1a). The adjustable capacity coupler of the surfatron and a triple coaxial stub allowed microwave power to be coupled to the discharge with a maximum of 5% reflected power.

Gas mixtures of Ar-N_2 with a N_2 maximum content of 1% were obtained from high-purity (>99.999%, Abello Linde S.A., Seville, Spain) Ar and N_2 gases, whose flows were controlled by mass flow controllers (HI-TEC, Bronkhorst, Iberfluid Instruments S.A., Barcelona, Spain). The total gas flow used in all experiments was kept at 1 slm (standard liter per minute).

When nitrogen takes part in the plasma gas mixture, two changes take place in the morphology of the plasma. Firstly, the discharge expands radially due to the higher thermal conductivity of the mixed plasma gas relative to that in pure Ar gas [16]. Due to the high temperatures and the presence of active species, this expansion can erode the tube containing the plasma and incorporate the ablated material to the discharge, thus the maximum nitrogen amount considered in the current work was chosen to ensure the integrity of quartz tube containing the discharge [17] and avoiding the contamination of plasma gas. Secondly, the addition to a pure Ar discharge, even in small amounts, leads to a noticeable reduction of plasma column length and the appearance of a postdischarge containing active species and very few or no charged particles, where neutral excited species with long lifetimes control its internal kinetics [18,19]. The experimental conditions considered in the current work ensured the obtention of a postdischarge long enough to act on the surface of the aluminum samples. Depending on the pressure conditions under which the plasma is generated, the nitrogen proportion in the Ar-N_2 mixture as well as the total plasma gas flow, two distinct regions can be observed in the postdischarge: (a) The early postdischarge (EAP), named pink afterglow as a results of the First Negative System (FNS) emission of N_2^+ in the 325–590 nm range, and (b) the late postdischarge (LAP) characterized by an orange color because of the emission of the First Positive System (FPS) of N_2 in the 500–1000 nm interval. In Figure 1b, a photograph corresponding to Ar-N_2 (1% N_2) used in the current investigation allows us to observe the plasma and the postdischarge; this last one expands in the air which leads to a reduction of its length in comparison to the postdischarges generated inside the same discharge tube in which the plasma is created [20,21]. This shortening could be due to the deexcitation of the excited neutral species of N_2, coming from the plasma, with the air surrounding the postdischarge.

The species generated into the plasma and the postdischarge were identified from the analysis of the radiation collected by an optical fibre of 1000 µm in diameter placed perpendicular to the tube axis (Figure 1a) and driven to the entrance slit of a monochromator (FHR-640, Jobin–Yvon Horiba, M.T. Brandao S.L., Madrid, Spain) type Czerny–Turner equipped with 2400 lines/mm holographic grating (200–750 nm range) and equipped with a Charge-Coupled Device (CCD) camera (Symphony, Jobin–Yvon Horiba, M.T. Brandao S.L., Madrid, Spain) as detector.

The aluminum samples (2.5 × 2.5 cm) were obtained from commercial aluminum coil whose surface composition was already analyzed in a previous work [14], mainly consisting of Al, O, C, and F. These samples were prepared following the procedure described in our previous work [14]: after cutting, samples were cleaned in an ultrasonic bath with acetone for 5 min and later they were dried in contact with air. Later, the samples were rinsed with ethanol and dried. This procedure is similar to that employed by other authors [7,12,13] and serves only to the purpose of removing the solid impurities on the material surface before the postdischarge treatment.

In order to determine the effective surface area treated by the postdischarge (spatial resolution), the wettability of the aluminum surfaces under the different conditions considered in the current study was measured from the water contact angle (WCA) of drops placed on the sample surfaces. For the measurement of the contact angle, the sessile drop method reported in [22] was used. Water drops of 5 µL (micro-liter) were deposited on aluminum surface after the treatment using a micropipette and pictures right after the deposition were taken by a Casio EXFH20 digital camera (Casio, Tokyo, Japan) with a focal distance of 5 mm and exposure time of 1/6 s (Figure 2). The values of the contact angle were obtained from three samples treated under the same experimental conditions and taken at different positions away from the tube axis ($x = 0$ mm), i.e., treatment axis (Figure 1a).

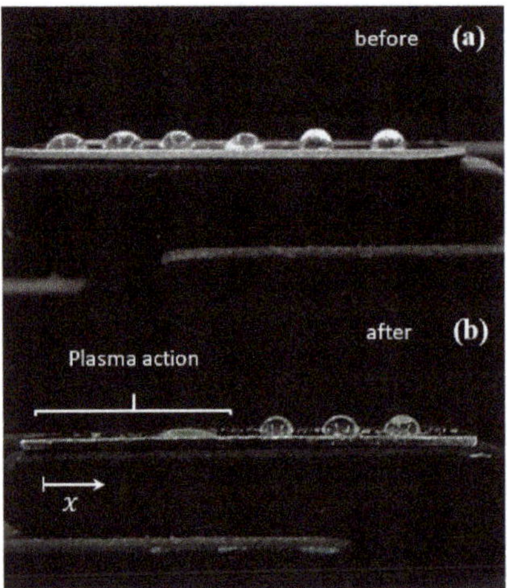

Figure 2. Photographs of water drops deposited in an aluminum sample before (**a**) and after the treatment (**b**) (1% N_2 in Ar-N_2, 150 W).

For their exposition to the postdischarge, the samples were placed on a two-dimensional displacement system formed by two linear actuators (T-LSM50B, Zaber Technologies Inc., Vancouver, BC, Canada); the movement of these actuators was controlled using a script executed in a Zaber console system. In Figure 1a, a red dash-line shows the movement of the platform on which the samples are placed. In order to obtain information about the spatial distribution of the proposed treatment, different experiments were carried out. In this way, the samples were placed at different distances (z) measured from the end of the discharge tube (Figure 1a), which matched the end of the discharge and, therefore, the beginning of the postdischarge. Due to the dimensions considered in this research, aluminum samples were only treated by the late postdischarge (see Figure 1b). Besides, the influence of the exposition time of the samples to the postdischarge was also studied as a function of the different speeds of the sample holder (v). For the purpose of ensuring the reproducibility of the treatment, different sets of experiments were carried out. In this sense, the resulting WCA profiles shown in Section 3.3 were built from the superposition of the measurements along the x direction, perpendicular to the treatment (Figure 1a), carried out in three different samples treated using the same conditions.

3. Results and Discussion

3.1. Active Species in the Discharge

The existence of the postdischarge area shows that Ar-N_2 discharges can be utilized to generate active species that can be used for remote surface treatment. Thus, a brief study of Ar-N_2 discharge by means of its emitted spectrum has been conducted to understand the formation of active species in the postdischarge.

Typical spectrum emitted by an Ar-N_2 (1%) discharge is shown in Figure 3, together with the spectrum emitted by a pure Ar discharge for comparison purposes. In pure Ar (Figure 3a), the spectrum is dominated by emissions belonging to the argon atomic system. Besides, only molecular bands corresponding to excited molecular radicals such as OH $\left(A^2\Sigma^+ \rightarrow X^2\Pi \right)$ and NH $\left(A^3\Pi \rightarrow X^3\Sigma^- \right)$ are observed. These radicals reveal the presence of water and nitrogen as impurities contained in the plasma gas since the discharge is contained in the quartz tube without being in contact with air. When N_2 is present in the plasma gas even in concentrations as low as 1% (Figure 3b), the spectral lines corresponding to 5p and upper levels of argon atomic system practically disappear and only few lines for 4p levels can be observed, and the Ar-N_2 spectrum is dominated by molecular bands from the radiation emitted by the nitrogen excited molecules and molecular ions (N_2^+).

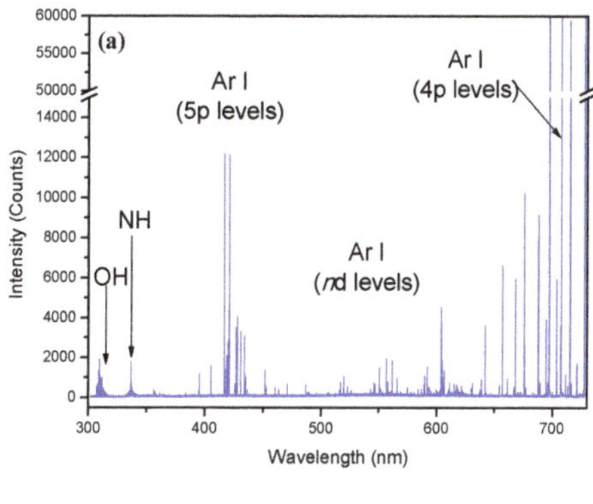

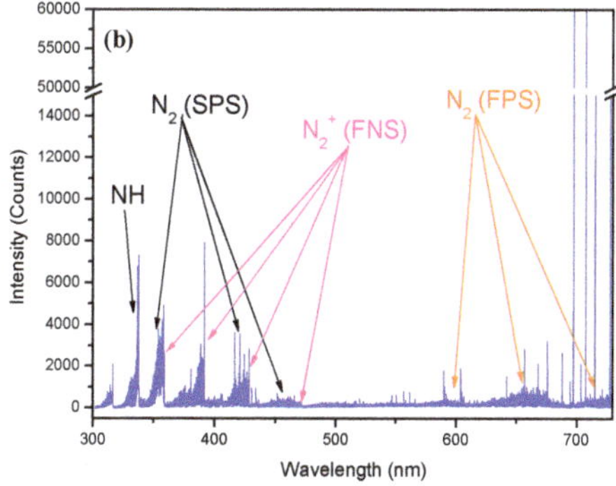

Figure 3. Optical emission spectra of a pure Ar (**a**) and an Ar-N_2 (1% N_2) (**b**) plasma.

In Figure 4, the main energy levels of Ar, N_2 molecule, and N_2^+ molecular ion are depicted; it can be seen that the mestastable levels of nitrogen, represented by $N_2\left(A^3\Sigma_u^+\right)$, have an energy equal to 6.17 eV which is lower than the mestastable level for Ar atom (11.5 eV). In plasmas generated at atmospheric pressure, the excitation/ionization processes are controlled by collisions with electrons, being carried out in steps, being the first excited levels (metastables) the departure levels for these processes [23–26]. This fact, together with the difference between the energies of metastable levels for Ar and N_2, induces a quick increase on the intensities of N_2 molecular and N_2^+ bands and significant changes in the emission of excited Ar atoms (Figure 3) when nitrogen is present. These differences are due to the competition of Ar metastable states and N_2 molecules at ground state for the plasma electrons of low energies (4–7 eV) (Figure 4), contributing to break the stepwise excitation chain of Ar atoms, which is reflected in the strong decrease in the radiation emission coming from the 5p and upper levels belonging to the Ar atomic system. This behavior is opposite to that found in Ar-He and Ar-Ne discharges, where the spectra are dominated by emissions of excited Ar atoms since the excitation of their metastable states requires collisions with electrons with energies higher than for Ar (19.8 and 16.6 eV for He and Ne, respectively) [27,28].

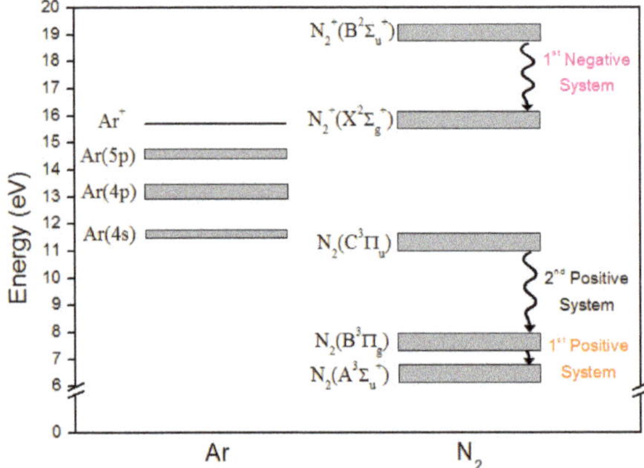

Figure 4. Main energy levels for the Ar atom and the N_2 molecule.

However, besides inelastic collisions with electrons, the formation of nitrogen excited species leading to the appearance of molecular bands in the spectra recorded for Ar-N_2 plasmas is also affected by processes involving heavy particles. Penning excitation (1), is the most probable for the excitation of nitrogen molecules up to $N_2(C^3\Pi_u)$ state [29]. This reaction is favored by the nearly resonant energies for these species; that is, 11.50 eV for Ar metastable atoms and 11.03 eV for $N_2(C^3\Pi_u)$ nitrogen excited molecules (Figure 4). From this level, spontaneous emission reactions (2) and (3), can also contribute to the formation of nitrogen molecules to excited states $N_2(B^3\Pi_g)$ and $N_2(A^3\Sigma_u^+)$, and the emission of the SPS and FPS, respectively.

$$Ar^m + N_2(X^1\Sigma_g^+) \rightarrow N_2(C^3\Pi_u) + Ar^0 \tag{1}$$

$$N_2(C^3\Pi_u) \rightarrow N_2(B^3\Pi_g) + h\nu(SPS) \tag{2}$$

$$N_2(B^3\Pi_g) \rightarrow N_2(A^3\Sigma_u^+) + h\nu(FPS) \tag{3}$$

Nitrogen molecular ions (N_2^+) are mostly created by charge transfer reactions (4) due to the resonance in energy of the particles taking part in this reaction [30], 15.6 eV for $N_2^+\left(X^2\Sigma_g^+\right)$ and 15.8 eV for Ar^+.

$$Ar^+ + N_2\left(X^1\Sigma_g^+\right) \rightarrow Ar^0 + N_2^+\left(X^2\Sigma_g^+\right) \tag{4}$$

The excited $N_2^+(B^2\Sigma_u^+)$, whose emission gives place to the first negative system (5) can be formed from the fundamental level of the molecular ion through inelastic collisions with electrons (6), with other nitrogen excited molecules ($N_2(A^3\Sigma_u^+)$, $N_2(B^3\Pi_g)$ and $N_2(C^3\Pi_u)$) taking part in the creation of fundamental level of ion molecular as intermediate steps.

$$N_2^+(B^2\Sigma_u^+) \rightarrow N_2^+\left(X^2\Sigma_g^+\right) + h\nu(FNS) \tag{5}$$

$$N_2^+\left(X^2\Sigma_g^+\right) + e^- \rightarrow N_2^+(B^2\Sigma_u^+) + e^- \tag{6}$$

Finally, formation of nitrogen atoms can be obtained from electron impact dissociation of N_2 molecules at ground state (7) or through processes of dissociative recombination of molecular ions (8) [31].

$$N_2\left(X^1\Sigma_g^+\right) + e^- \rightarrow N + N + e^- \tag{7}$$

$$N_2^+\left(X^2\Sigma_g^+\right) + e^- \rightarrow N + N \tag{8}$$

The presence of nitrogen atoms in the discharge is supported by the emission of the band originated by NH $\left(A^3\Pi\right)$ (Figure 3b), being NH formed by recombination of free nitrogen atoms and hydrogen in the discharge in the presence of a third body (9) [32]. In the case of pure Ar plasma (Figure 3a), the presence of NH bands in the spectrum is due to the nitrogen and hydrogen atoms present in the plasma as impurities.

$$N + H + Ar^0 \rightarrow NH\left(A^3\Pi\right) + Ar^0 \tag{9}$$

3.2. Active Species in the Postdischarge

In surface-wave plasmas, the discharge ends at the point where the wave sustaining the plasma ceases its propagation. Excited particles with long lifetimes, created in the discharge and draught downstream by the gas flow, can still exist outside of the discharge, giving place to the postdischarge. However, since no external power is supplied to the postdischarge, the existence of particles capable of interacting with the samples exposed to them depends entirely on the energy of these long-lived particles and their kinetics. Figure 5 shows the spectra corresponding to the radiation emitted by the postdischarge taken at different distances measured from the end of the discharge tube (plasma end). At $z = 0.5$ cm, closer the plasma end, OH, NH, N_2, and $N_2{}^+$ molecular transitions (Figure 5a) together with the emissions from the ArI atomic system (Figure 5) can be observed. As we move along the postdischarge, all signals decrease, thus implying a decrease in the concentration of active species.

In particular, the OH band signal can be appreciated at 0.5 cm from the end of the discharge, but it disappears completely at a distance of 3 cm. This fact suggests that no new OH radicals are generated in the postdischarge and mission of the NH band in the postdischarge points out the presence of nitrogen atoms in this region. A decrease of the intensity of this band is related to the reduction of the numthus, its presence is due to transport of this specie by the gas flow from the discharge but not to the interaction with the air surrounding the postdischarge. According to the formation pathway of NH through Equation (9), the eber of N atoms while moving further the end of the discharge. Indeed, from Figure 5a, the presence of N atoms is demonstrated at postdischarge positions of $z = 3$ cm due to the emission of NH band.

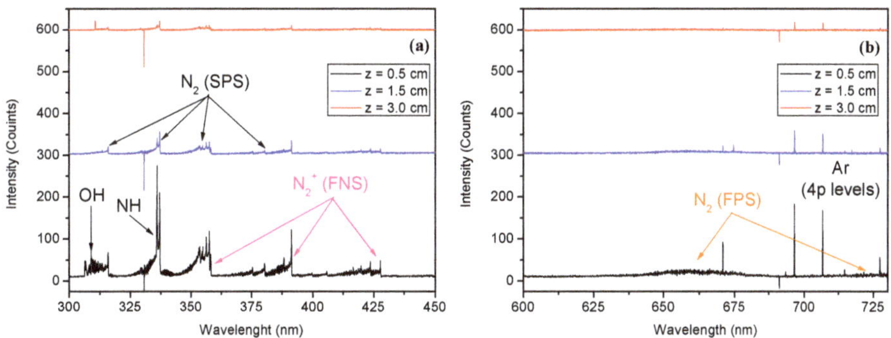

Figure 5. Optical emission spectra of the postdischarge at different axial positions away from the end of the discharge in the (**a**) 300–450 nm and (**b**) 600–730 nm spectral ranges.

The Second Positive System (SPS), whose signal is due to the existence of $N_2\left(C^3\Pi_u\right)$ excited molecules could either be generated by Penning excitation (1) with Ar metastable atoms or by nitrogen metastable pooling (10). Both formation channels are possible since the existence of both Ar metastable atoms and nitrogen $N_2\left(A^3\Sigma_u^+\right)$ metastable molecules in the postdischarge region is demonstrated by the emission of Ar lines and the FPS, respectively.

$$N_2\left(A^3\Sigma_u^+\right) + N_2\left(A^3\Sigma_u^+\right) \rightarrow N_2\left(B^3\Pi_g, C^3\Pi_u\right) + N_2\left(X^1\Sigma_g^+\right) \tag{10}$$

The existence of $N_2\left(B^3\Pi_g\right)$ excited molecules is linked, besides metastable pooling, to the three-body recombination for the nitrogen atoms (11)

$$N + N + Ar^0 \rightarrow N_2\left(B^3\Pi_g\right) + Ar^0 \tag{11}$$

and followed by deexcitation by reaction (3) gives place to the FPS, whose presence indicates the existence of metastable $N_2\left(A^3\Sigma_u^+\right)$ molecules in the postdischarge.

Finally, given that no additional energy is supplied to the postdischarge, very few electrons can be created and most of the electrons coming from the discharge are readily lost by recombination. However, the emission of the FNS of the nitrogen molecular ion can still be detected at $z = 3$ cm.

3.3. Spatial Distribution of the Plasma Action on the Surface of Aluminum Samples

Changes in the WCA measured following the procedure described in Section 2 can provide information about the wettability modifications suffered by the material exposed to the postdischarge. As it has been previously described, treatment profiles were built from the WCAs measured for three different samples which have been plotted with different colors in the following figures (Figures 6, 8 and 9). The resulting profiles were fit to a simple sigmoid function (12)

$$WCA = WCA_{min} + \frac{WCA_{max} - WCA_{min}}{1 - e^{\frac{x - x_0}{dx}}} \tag{12}$$

being WCA_{min} and WCA_{max} the minimum and maximum values of the contact angle, respectively, x_0 the inflection point of the sigmoid function, which can be taken as a measurement of the radius of the treatment spot. Finally, dx is a parameter related to the width of the transition zone between WCA_{min} and WCA_{max}, which in our case was fixed to 0.5 mm for all the profiles as it provided the best fitting results.

In Figure 6, the WCA profile obtained at different positions away from the treated zone for aluminum samples treated with the postdischarges of an Ar and A-N_2 (1%) plasmas appear depicted. For these experiments, samples were located at $z = 4$ cm away from the end of the discharge tube and the displacement speed was maintained constant at 7500 µm/s. The similarity of the profiles obtained for different samples is a prove of the reproducibility of the treatment. The resulting dispersion in WCA values can be explained since all the samples were obtained from commercial aluminum, which can contain imperfections resulting from machining affecting the WCA measurement. To illustrate these imperfections, microscopic images from the samples were taken using a JEOL (Tokyo, Japan) JSM 7800 Prime scanning electron microscope (SEM) (Figure 7). There, striped marks resulting from machining as well as other imperfections are noticed both in the untreated and treated samples. No significant difference in the morphology nor smoothing of the surface can be observed for the case of the treated samples.

From Figure 6, a significant change in the WCA is observed when nitrogen is present in the postdischarge, as compared to the case of the pure Ar postdischarge. In this way, exposing the surface to the postdischarge of a plasma containing 1% N_2 results in an improvement of the wettability properties of the aluminum surfaces showing a decrease of WCA_{min} from 80° to 40°. Studies carried out at atmospheric pressure have suggest that the light emitted by the discharge, containing high-energy photons, might play some role in breaking organic bonds at the surface of polymeric materials [33]. However, our results points to an important role of the nitrogen species in the modification of the wettability of the surfaces. Indeed, similar studies performed at reduced pressures have suggested that nitrogen atoms can play a significant role in the cleaning process of iron foils, where aliphatic carbon atoms existing in the surface would be removed after association with nitrogen atoms as CN radicals [34], which is a more likely explanation given that the Ar postdischarge does not produce a significant effect on the surface (Figure 6a). Furthermore, this mechanism is in good agreement with the results observed for aluminum samples treated with similar procedures (Figure 7), where

X-ray photoelectron spectrometry (XPS) showed an increase in the amount of highly oxidized forms of nitrogen at the surface after the treatment with Ar-N_2 postdischarges as well as the appearance of hydroxide groups [14].

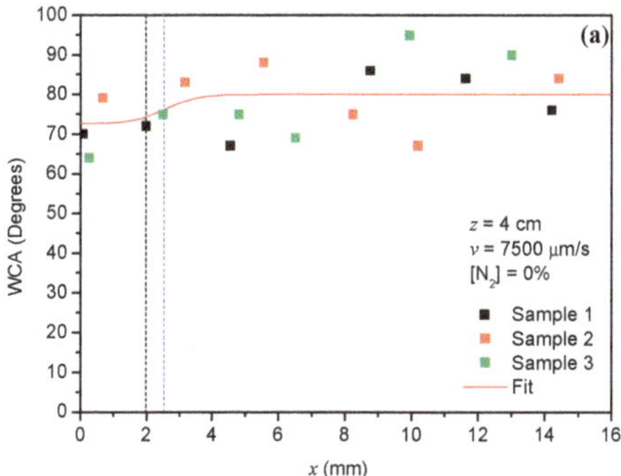

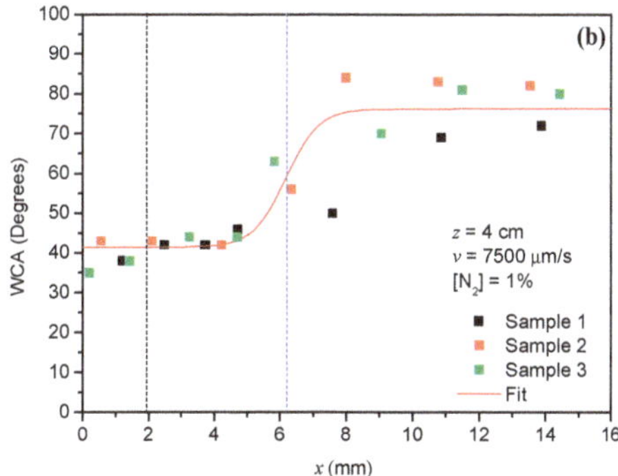

Figure 6. Water contact angle profiles after (**a**) Ar and (**b**) Ar-N_2 postdischarge treatment.

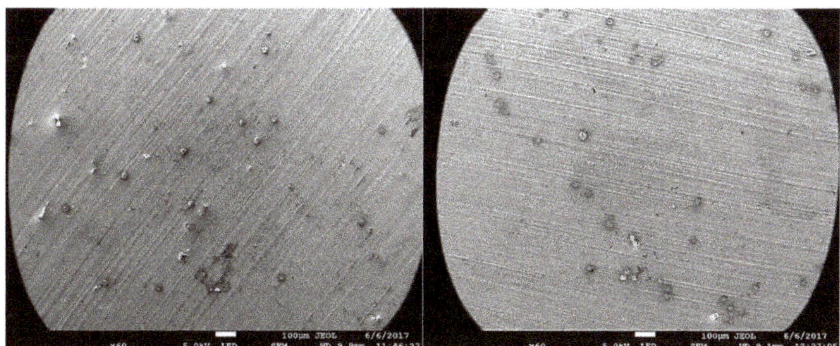

Figure 7. Scanning electron microscope image of aluminum samples before (**left**) and after (**right**) the treatment with an Ar-N_2 (1% N_2) postdischarge.

Moreover, in Figure 6b, the radius of the treated area has also been marked by a vertical blue dashed line, together with the inner radius of the tube containing the discharge, whose size appears represented as a vertical dashed black line for comparison purposes. For the considered conditions, the radius of the treated zone is of 6.1 ± 0.4 mm, which is three times the inner radius of the tube containing the discharge, thus meaning that the postdischarge expands before reaching the sample, and might be able to interact with the surrounding atmosphere.

The influence of varying the treatment distance, z, for the same speed and nitrogen concentration can be observed comparing Figures 8 and 6b. For increasing treatment distances from 2 to 6 cm, WCA_{min} rising from values of 25° to 65°, respectively, revealing a decrease in the cleaning effect of the postdischarge as its length increases. This fact agrees with the proposed mechanism of action mainly due to the active species of postdischarge. When the distance from the end of the discharge increases, the amount nitrogen excited species decreases and, consequently, the capability of the postdischarge to modify the hydrophobicity of the material surface. These results are in agreement with those reported for similar treatments [14].

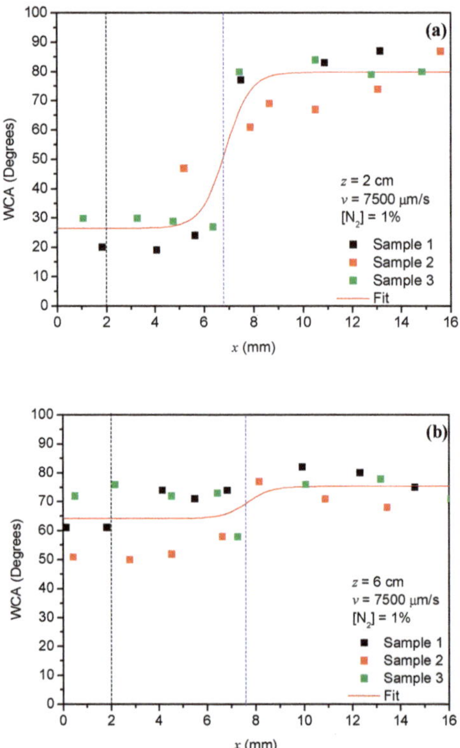

Figure 8. Water contact angle profile after Ar-N$_2$ postdischarge treatment at different distances measured from the end of the discharge: (**a**) $z = 2$ cm and (**b**) $z = 6$ cm.

The width of the treatment profile is not significantly influenced by the distance of the treatment, showing a radii of 6.9 ± 0.3 and 7.7 ± 1.3 mm for distances of 2 and 6 cm, respectively. This indicates that, for the considered conditions, the postdischarge expands once the gas exits the tube containing the discharge within a relatively short distance (≤2 cm). From this point, the transversal section maintains practically a radius value around 6–7 mm. This happens because the outer layers of the gas shield the inner part of the postdischarge and reduce the interaction between the postdischarge and the surrounding atmosphere. A similar behavior has been described for plasma gas used to generate a microwave plasma torch [31].

The effect of treating the samples at different treatment speeds; i.e., at different treatment times, while keeping the distance at 4 cm is shown in Figure 9. Comparing with the results obtained in Figure 6b, reducing the speed of the treatment to 5000 µm/s does not significantly affect the effect of the treatment, resulting in WCA_{min} values of 45°, and a radius of the treatment spot of 5.8 ± 0.3 mm, similar to those obtained for a treatment speed of 7500 µm/s. Further reduction of the treatment speed to 500 µm/s leads to a radius of the treatment spot of 4.4 ± 1.1 mm, which reasonably agrees with the previous ones. The most significant difference in the case when the sample is treated for a longer time, i.e., lower treatment velocity, is that WCA_{max} decreases to values around 60°. It might be likely a result of heating of the area adjacent to the treatment spot and the subsequent vaporization of hydrocarbons present at the surface. Even though the effect of heating was discarded in previous studies for speeds of 7500 µm/s [14], it may still play a role for lower speeds. In any case, it must be taken into account that, for this case, dispersion in the WCA results is larger than in the previous ones.

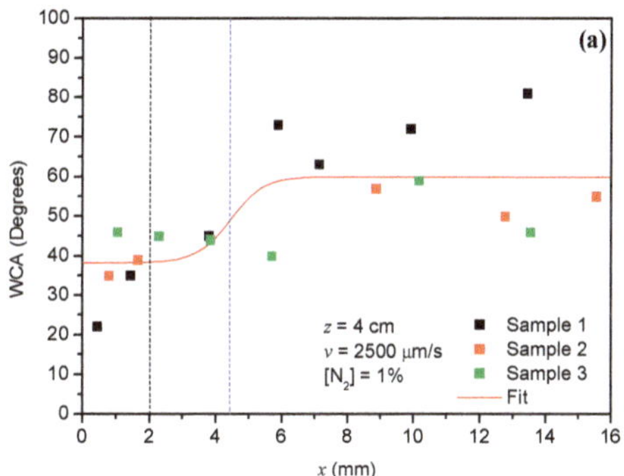

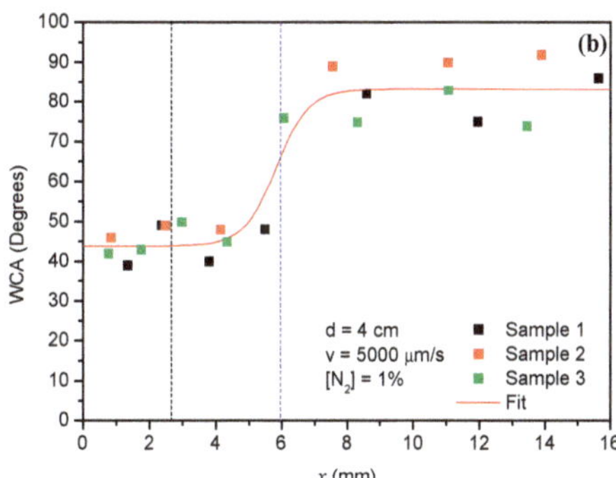

Figure 9. Water contact angle profile after Ar-N_2 postdischarge treatment at different speeds: (**a**) 2500 µm/s and (**b**) 5000 µm/s.

4. Conclusions

The parameters influencing the spatial distribution of the treatment of an aluminum surface by an atmospheric pressure remote plasma have been investigated using the sessile drop method. The position of the sample in the postdischarge (z) has been found to have a large impact in the wettability property of the treated sample, with lower WCA for positions closer to the end of the discharge. On the other hand, the treatment speed (v) has been shown to have a little impact on the spatial distribution of the treatment, except for the lowest velocity, where a lowering in the WCA values of the outer zone of the profile has been found, probably due to heating of the adjacent zones.

Neither the distance to the plasma end (z) nor the treatment speed (v) have a significant impact on the radius of the spot, which remains approximately constant (6–7 mm) regardless of the considered conditions (2 < z < 6 cm, 2500 < v < 7500 µm/s).

The possible cleaning and activation mechanism was discussed to the light of the spectra emitted by the postdischarge, where active species with long lifetimes are draught from the discharge by the action of the gas flow. The limited effect of pure argon postdischarges on the surface for the treating conditions considered suggests that active species of nitrogen are the main agent responsible for the action of the postdischarge on the aluminum surfaces. Due to the absence of significant OH emissions at relatively short distances, the direct implantation of these radicals might be discarded, suggesting that interaction with other active and/or energetic species in the discharge is responsible for the increase of the surface wettability of the treated samples.

Author Contributions: J.M. designed and performed the experiments, M.D.C. analyzed data, J.M., R.R. and M.D.C. wrote and reviewed the paper.

Funding: This research was funded by Universidad de Cordoba (Spain), grant number MOD 4.1 PP2017-2018 (XXII-XXIII Programa Propio de Fomento de la Investigación).

Acknowledgments: This work was supported by XXII-XXIII Programa Propio de Fomento de la Investigación de la Universidad de Córdoba (Spain) (MOD 4.1 PP2017-2018).

Conflicts of Interest: The authors declare no conflict of interest.

References

1. Richard Prabakar, S.J.; Hwang, Y.-H.; Bae, E.G.; Lee, D.K.; Pyo, M. Graphene oxide as a corrosion inhibitor for the aluminum current collector in lithium ion batteries. *Carbon* **2013**, *52*, 128–136. [CrossRef]

2. Liu, Y.; Zhang, J.; Li, S.; Wang, Y.; Hana, Z.; Rena, L. Fabrication of a superhydrophobic graphene surface with excellent mechanical abrasion and corrosion resistance on an aluminum alloy substrate. *RSC Adv.* **2014**, *4*, 45389. [CrossRef]

3. Polini, W.; Sorrentino, L. Improving the wettability of 2024 aluminum alloy by means of cold plasma treatment. *Appl. Surf. Sci.* **2003**, *214*, 1–4. [CrossRef]

4. Morent, R.; de Geyter, N.; Verschuren, J.; de Clerck, K.; Kiekens, P.; Leis, C. Non-thermal plasma treatment of textiles. *Surf. Coat. Technol.* **2008**, *202*, 3427–3449. [CrossRef]

5. Xu, X. Dielectric barrier-discharge-properties and applications. *Thin Solid Films* **2001**, *390*, 237–242. [CrossRef]

6. Prysiazhnyi, V.; Matoušek, J.; Černák, M. Steel surface treatment and following aging effect after coplanar barrier discharge plasma in air, nitrogen and oxygen. *Chem. Listy* **2012**, *106*, 1475–1481.

7. Prysiazhnyi, V. Atmospheric pressure plasma treatment and following aging effect of chromium surfaces. *J. Surf. Eng. Mater. Adv. Technol.* **2013**, *3*, 138–145. [CrossRef]

8. Bónová, I.; Zahoranová, A.; Kováčik, D.; Zahoran, M.; Mičušík, M.; Černák, M. Atmospheric pressure plasma treatment of flat aluminum surface. *App. Surf. Sci.* **2015**, *331*, 79–86. [CrossRef]

9. Prysiazhnyi, V.; Zaporojchenko, V.; Kersten, H.; Černák, M. Influence of humidity on atmospheric pressure air plasma treatment of aluminum surface. *Appl. Surf. Sci.* **2012**, *258*, 5467–5471. [CrossRef]

10. Prysiazhnyi, V.; Vasina, P.; Panyala, N.R.; Havel, J.; Černák, M. Air DCSBD plasma treatment of Al surface at atmospheric pressure. *Surf. Coat. Technol.* **2012**, *206*, 3011–3016. [CrossRef]

11. Yamamoto, T.; Yoshizaki, A.; Kuroki, T.; Okubo, M. Aluminum surface treatment using three different plasma-assisted dry chemical processes. *IEEE Trans. Ind. Appl.* **2004**, *40*, 1220–1225. [CrossRef]

12. Hnilica, J.; Kudrle, V.; Potočnaková, L. Surface treatment by atmospheric-pressure surfatron jet. *IEEE Trans. Plasma Sci.* **2012**, *40*, 2925–2930. [CrossRef]

13. Shin, D.H.; Bang, C.U.; Kim, J.H.; Hong, Y.C.; Uhm, H.S.; Park, D.K.; Him, J.K. Treatment of metal surface by atmospheric microwave plasma jet. *IEEE Trans. Plasma Sci.* **2006**, *34*, 1241–1246. [CrossRef]

14. Muñoz, J.; Bravo, J.A.; Calzada, M.D. Aluminum metal surface cleaning and activation by atmospheric-pressure remote plasma. *Appl. Surf. Sci.* **2017**, *407*, 72. [CrossRef]

15. Moisan, M.; Zakrzewski, Z.; Pantel, R. The theory and characteristics of an efficient surface wave launcher (surfatron) producing long plasma columns. *J. Phy. D: Appl. Phys.* **1979**, *12*, 219–237. [CrossRef]

16. Kabouzi, Y.; Calzada, M.D.; Moisan, M.; Tran, K.C.; Trassy, C. Radial contraction of microwave sustained plasma columns at atmospheric pressure. *J. Appl. Phys.* **2002**, *91*, 1008–1019. [CrossRef]

17. Bravo, J.A.; Muñoz, J.; Sáez, M.; Calzada, M.D. Atmospheric pressure Ar-N_2 surface-wave discharge morphology. *IEEE Trans. Plasma Sci.* **2011**, *39*, 2114–2115. [CrossRef]

18. Callede, G.; Deschamps, J.; Godert, J.L.; Ricard, A. Active nitrogen atom in an atmospheric pressure flowing Ar-N_2 microwave discharge. *J. Phys. D: Appl. Phys.* **1991**, *24*, 909–914. [CrossRef]

19. Ricard, A.; Tétreault, J.; Hubert, J. Nitrogen atom recombination in high Ar-N_2 flowing post-discharges. *J. Phy. B: At. Mol. Opt. Phy.* **1991**, *24*, 1115–1123. [CrossRef]

20. Bravo, J.A.; Rincón, R.; Muñoz, J.; Sanchez, A.; Calzada, M.D. Spectroscopic characterization of argon-nitrogen Surface-wave discharges in dielectric tubes at atmospheric pressure. *Plasma Chem. Plasma Process.* **2015**, *35*, 993–1014. [CrossRef]

21. Rincón, R.; Yubero, C.; Calzada, M.D.; Moyano, L.; Zea, L. Plasma technology as a new food preservation technique. In *Microbial Food Safety and Preservation Techniques*; Ravishankar Rai, V., Bai, J.A., Eds.; CRC Press: Boca Raton, FL, USA, 2015.

22. Yuan, Y. Contact Angle and Wetting Properties. In *Surface Science Techniques*; Bracco, G., Holst, B., Eds.; Springer: Berlin, Germany, 2013.

23. Calzada, M.D.; Garcia, M.C.; Luque, J.M.; Santiago, I. Influence of the thermodynamic equilibrium state in the excitation of samples by a plasma at atmospheric pressure. *J. Appl. Phy.* **2002**, *92*, 2269. [CrossRef]

24. Kudela, J.; Odrobina, I.; Kando, M. High-speed camera study of the surface wave discharge propagation in xenon. *Jpn. J. Appl. Phy.* **1998**, *37*, 4169. [CrossRef]

25. Calzada, M.D.; Moisan, M.; Gamero, A.; Sola, A. Experimental investigation and characterization of the departure from local thermodynamic equilibrium along a surface-wave-sustained discharge at atmospheric pressure. *J. Appl. Phy.* **1996**, *80*, 46. [CrossRef]

26. Jonkers, J.; de Regt, J.M.; van der Mullen, J.A.M.; Vos, H.P.C.; de Groote, F.P.; Timmermans, E.A.H. On the electron temperatures and densities produced by the "torche à injection axiale". *Spectrochim. Acta Part B* **1996**, *51*, 1385. [CrossRef]

27. Muñoz, J.; Calzada, M.D. Experimental study on equilibrium deviations in atmospheric pressure argon/helium surface wave discharges. *Spectrochim. Acta Part B* **2010**, *65*, 1014–1021. [CrossRef]

28. Muñoz, J.; Rincón, R.; Melero, C.; Dimitrijević, M.S.; González, C.; Calzada, M.D. Validation of the van der Waals broadening method for the determination of gas temperature in microwave discharges sustained in argon–neon mixtures. *J. Quant. Spectrosc. Radiat. Transfer* **2018**, *206*, 135–141. [CrossRef]

29. Ricard, A. *Reactive Plasmas*; Société Française du Vide: Paris, France, 1996.

30. Henriques, J.; Tatarova, E.; Guerra, V.; Ferreira, C.M. Wave driven N_2-Ar discharge. I. Self-consistent theoretical model. *J. Appl. Phys.* **2001**, *91*, 5632–5639. [CrossRef]

31. Guerra, V.; Tatarova, E.; Dias, F.M.; Ferreira, C.M. On the self-consistent modeling of a traveling wave sustained nitrogen discharge. *J. Appl. Phys.* **2002**, *91*, 2648–2661. [CrossRef]

32. Rincón, R.; Muñoz, J.; Sáez, M.; Calzada, M.D. Spectroscopic characterization of atmospheric pressure argon plasmas sustained with the Torche à Injection Axiale sur Guide d'Ondes. *Spectrochim. Acta, Part B* **2013**, *81*, 26–35. [CrossRef]

33. Szili, E.J.; Al-Bataineh, S.A.; Bryant, P.M.; Short, R.D.; Bradley, J.W.; Steele, D.A. Controlling the spatial distribution of polymer surface treatments using atmospheric-pressure microplasma jets. *Plasma Processes Polym.* **2011**, *8*, 38–50. [CrossRef]

34. Mézerette, D.; Belmonte, T.; Hugon, R.; Henrion, G.; Czerwiec, T.; Michel, H. Study of the surface mechanisms in an Ar-N_2 post-discharge cleaning process. *Surf. Coat. Technol.* **2003**, *169–170*, 181–185. [CrossRef]

 *metals* 

Article

Comparison of Mechanical and Antibacterial Properties of TiO₂/Ag Ceramics and Ti6Al4V-TiO₂/Ag Composite Materials Using Combined SLM-SPS Techniques

Ramin Rahmani [1,*], Merilin Rosenberg [2,3], Angela Ivask [3] and Lauri Kollo [1]

[1] Department of Mechanical and Industrial Engineering, Tallinn University of Technology, Ehitajate Tee 5, 19086 Tallinn, Estonia
[2] Department of Chemistry and Biotechnology, Tallinn University of Technology, Ehitajate Tee 5, 19086 Tallinn, Estonia
[3] Laboratory of Environmental Toxicology, National Institute of Chemical Physics and Biophysics, Akadeemia Tee 23, 12618 Tallinn, Estonia
* Correspondence: ramin.rahmaniahranjani@ttu.ee; Tel.: +372-5198-8504

Received: 16 July 2019; Accepted: 7 August 2019; Published: 8 August 2019

Abstract: In present work, the combination of spark plasma sintering (SPS) and selective laser melting (SLM) techniques was introduced to produce composite materials where silver-doped titania (TiO₂) ceramics were reinforced with ordered lattice structures of titanium alloy Ti6Al4V. The objective was to create bulk materials with an ordered hierarchical design that were expected to exhibit improved mechanical properties along with an antibacterial effect. The prepared composite materials were evaluated for structural integrity and mechanical properties as well as for antibacterial activity towards *Escherichia coli*. The developed titanium–silver/titania hybrids showed increased damage tolerance and ultimate strength when compared to ceramics without metal reinforcement. However, compared with titania/silver ceramics alone that exhibited significant antibacterial effect, titanium-reinforced ceramics showed significantly reduced antibacterial effect. Thus, to obtain antibacterial materials with increased strength, the composition of metal should either be modified, or covered with antibacterial ceramics. Our results indicated that the used method is a feasible route for adding ceramic reinforcement to 3D printed metal alloys.

Keywords: Ti6Al4V lattice structure; Ag-doped TiO₂ anatase; spark plasma sintering; selective laser melting; additive manufacturing; antibacterial and photoactivity applications

1. Introduction

Titania (TiO₂) is one of the very often used ceramic materials in medical, antibacterial, paint, varnish, and pigment applications. Two well-known mineral forms of TiO_2 are anatase and rutile that have motivated interests in electrical conductivity and photocatalytic activity fields. Anatase TiO_2 has been shown to exhibit photocatalytic and thus, self-cleaning, activity under ultraviolet illumination. Photodegradation capability of TiO_2, especially under visible light conditions can be even further enhanced by depositing transition metal dopants like silver (Ag) [1,2]. For enhanced antibacterial effect of those materials, Ti-implants have been coated by TiO_2-nanotubes and Ag-nanoparticles in dental and orthopedic applications [3]. The primary interest in TiO_2 ceramics has been related to its use in thin films or as an additive [4]. The implementation of TiO_2 as a bulk ceramic is mainly restricted due to its high brittleness, having fracture toughness in the order of 2–3 MPa·m$^{1/2}$ [5].

It is possible to reinforce ceramic materials with metals, to improve the mechanical properties of the ceramics. Titanium and its alloys are widely used as biomaterials due to their sufficient biocompatibility, light weight, and high mechanical strength [6]. Titanium has excellent physical properties, its corrosion resistance is utterly known for orthopedic, osteology, and dental applications [7] and it possesses the ability to be 3D printed into complex objects. Research studies have displayed that the powder of Ti6Al4V, a material that is finding increasing use in medical applications has a high potential as a starting material for selective laser melting (SLM), a method that can be used to create various 3D structures [8]. Titania that requires relatively low temperatures for full consolidation, may be successfully combined with 3D printed metals. Combining oxide ceramics with metals by common methods is however complicated due to the inherent incompatibility of the interphases. For example, fusing of wear resistant oxide ceramics on a titanium substrate for implants faces distinct challenges in terms of obtaining strong bonding between the ceramic and the metal [9].

Different methods can be applied for consolidating ceramics and composites based on these. When ultrafine- or nano-structures are desired, spark plasma sintering (SPS) is commonly used. Both nanostructured titania and titania based nanocomposites have been produced by SPS [10–12]. Due to very high heating and cooling rate that can preserve the nanostructure at nearly full density, silver-doped titania could be employed for medical applications where antimicrobial properties, chemical inertness, and wear resistance are needed and also for water purification. Combining with titanium could drastically increase the damage tolerance of the ceramic and also provide added application specific functions. For example, a device of combined TiO_2/Ag and Cu/CoNiP was shown to effectively perform as magnetically rolling microrobots for water purification [13].

In this study a combination of selective laser melting and spark plasma sintering was introduced for the production of new versatile metal–ceramic hybrid structures to increase damage tolerance of the material. In the process, SLM was used to first 3D print the periodic titanium lattice structure followed by the embedding of titania–silver composite powder to the lattice and hot consolidation by SPS. Prepared materials were analyzed for mechanical properties under a compression test and antibacterial activity against *Escherichia coli* cells.

2. Materials and Methods

2.1. Materials

Titanium dioxide powders with BET surface area of 150 m^2/g (Figure 1a) and flaky silver powder with the purity of 99.95% (Figure 1b) were purchased from ABCR GmbH and used to produce ceramic materials and as the matrix phase in composites. Titanium dioxide was first ultrasonically deagglomerated under isopropanol. Doping with 2.5wt% of silver was performed by using a bottle mixer at 15 RPM for 24 h. Yttria (Y_2O_3) stabilized polycrystalline zirconia (ZrO_2) was manufactured in TOSOH corporation (TZ-3Y-E, Tokyo, Japan) and was used to produce control ceramic surfaces for antibacterial tests (Figure 1c). Gas atomized Ti6Al4V alloy powders having a particle size in the range from 15 to 45 µm (Figure 1d) were obtained from TLS Technik Spezialpulver GmbH and were used to print metal lattices.

2.2. Specimens Preparation

Ceramic materials of TiO_2–2.5% Ag, pure TiO_2 and ZrO_2 (the latter two as controls for the antibacterial test) were produced using a spark plasma sintering (SPS) machine HP D10 from FCT GmbH. Sintering temperature of 750 °C, pressure of 75 MPa and holding time of 30 min were used to compact the composites. To produce metal–ceramic composite materials, first cylindrically shaped Ti6Al4V lattices (Figure 2, top row) were produced using selective laser melting (SLM50 metal additive manufacturing system from Realizer GmbH). The specimens had a diameter of 20 mm and height of 15 mm. The specimens exhibited diamond type porous lattices with unit cell size from 1 to 2 mm. To prepare metal–ceramic composites, porous titanium lattice was placed in a graphite mold and ceramic powder of TiO$_2$ supplemented with 2.5% Ag was embedded in the lattices (Figure 2, bottom row). SPS at 900 °C, 75 MPa and during 30 min was used to compact the composites (Figure 3). The height of compacted composite lattices was 3–4 mm and depended on the cell size of initial lattices. Lattices with 1 mm cell size were shrunk down to 4 mm, lattices with 1.5 mm cell size shrunk to 3.5 mm and lattices with 2 mm cell unit size shrunk to 3 mm. For the compressive test, identical SPS conditions were applied for 10 mm diameter and 25 mm height lattice structures.

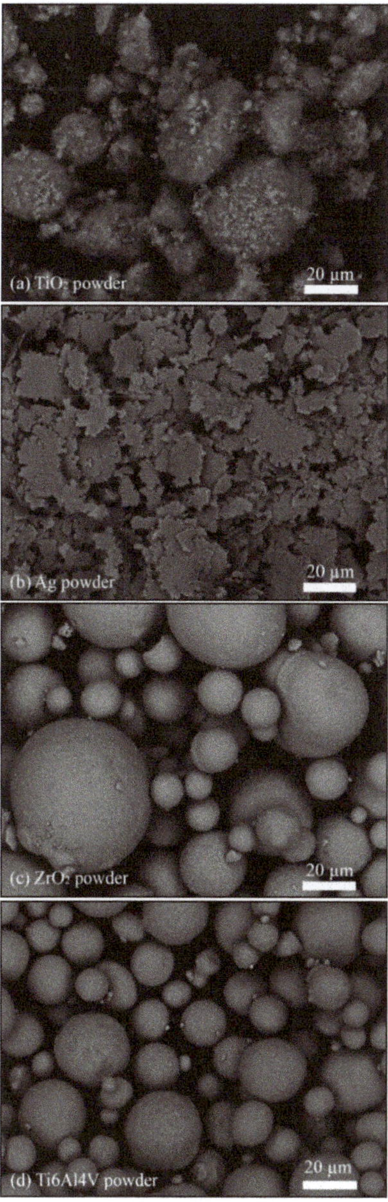

Figure 1. SEM micrographs of (**a**) TiO$_2$ anatase, (**b**) Ag, (**c**) ZrO$_2$, and (**d**) Ti6Al4V powders.

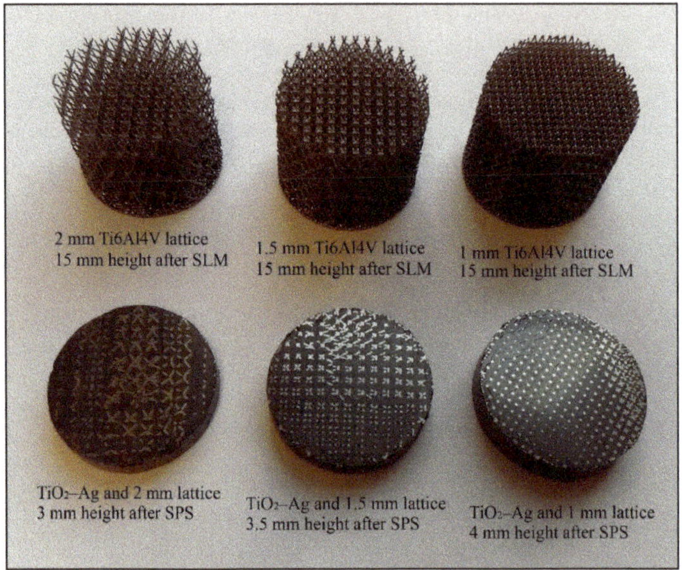

Figure 2. Top row: Selective laser melting (SLM) manufactured Ti6Al4V lattice structures with unit cell size and volume fraction of 2 mm and 6%, 1.5 mm and 9%, 1 mm and 16%, respectively (dimensions of lattice structures are 20 mm diameter and 15 mm height); **Bottom row**: Spark plasma sintering (SPS) sintered TiO_2–2.5% Ag embedded in the Ti6Al4V lattice structures.

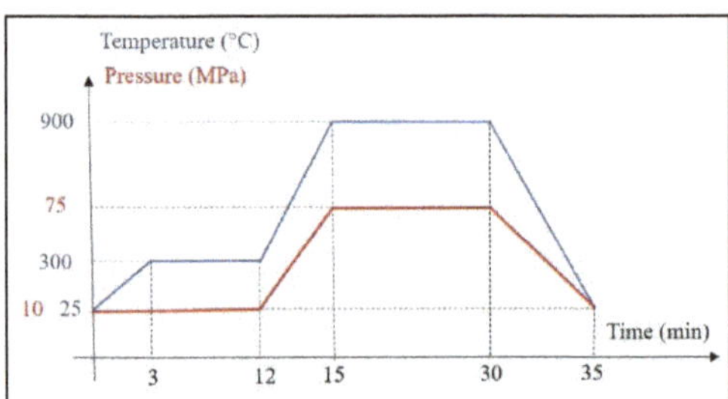

Figure 3. SPS conditions for TiO_2 powder embedded Ti6Al4V lattice structure.

2.3. Mechanical and Microstructural Characterization

Microstructure of the produced materials was examined with a scanning electron microscope (Zeiss EVO MA15, Oberkochen, Germany) equipped with energy dispersive spectroscopy (EDS). Compressive testing of the samples was performed on Instron 8516 servo-hydraulic test machine. The cylindrical samples with diameter of 10 mm and height of 8–9 mm (the height of 3D printed lattice structures was 25 mm which was shrunken to 8–9 mm after SPS) were loaded with a crosshead speed of 0.5 mm/min, according to standard ASTM E9/09.

2.4. Antibacterial Assay

A comparative antibacterial assay was carried out for TiO_2–2.5% Ag, TiO_2, and ZrO_2 ceramics as well as for TiO_2–2.5% Ag and Ti6Al4V lattice hybrid structures of 1, 1.5, and 2 mm cell sizes (Figure 2). The assay was carried out using an in-house protocol based on ISO 27447:2009 and ISO 22196:2007 standard methods [14] towards a model gram-negative bacterium *Escherichia coli* MG1655. Prior to all experiments, the specimens were sanded and polished, in order to remove potential contaminants and smoothen the surface, then sterilized by autoclaving at 121 °C for 15 min. The material samples were reused for consecutive experiments. After each test, samples were thoroughly washed with water and 70% ethanol, drained, submerged in 80 mL deionized water and sonicated using Branson Digital Sonifier model 450 (max power 400 W) equipped with horn model 101-135-066R at 25% amplitude for 10 min before autoclaving. *E. coli* culture for inoculum suspension was collected from fresh nutrient agar (5 g/L meat extract, 10 g/L peptone, 5 g/L sodium chloride, 15 g/L agar powder in deionized water) plates incubated overnight at 30 °C, suspended in 500-fold diluted nutrient broth (3 g/L meat extract, 10 g/L peptone, 5 g/L sodium chloride in deionized water) and further diluted with the same medium to optical density of 0.01 at 600 nm. Sterile surface samples were placed on the bottom of sterile 6-well polystyrene plates, inoculated with 50 µL *Escherichia coli* MG1655 suspension and covered with 2 cm × 2 cm × 0.005 cm polyethylene film. Exposure medium was 1:500 diluted nutrient broth. Samples were in parallel either covered by 1.1 mm UVA-transmissive borosilicate glass sheet and exposed to 2–2.5 W/m^2 UVA at 315–400 nm spectral range (measured using Delta Ohm UVA probe) effective at the sample level or kept in the dark covered by a 6-well plate lid.

After 30 min and 4 h exposure bacteria were retrieved from samples by repeatedly pipetting 3 mL of neutralizing medium (soybean-casein digest broth with lecithin and polyoxyethylene sorbitan monooleate: 17 g/L casein peptone, 3 g/L soybean peptone, 5 g/L sodium chloride, 2.5 g/L disodium hydrogen phosphate, 2.5 g/L glucose, 1.0 g/L lecithin, 7 g/L nonionic surfactant in deionized water) over the surface, serially diluted in 2 mL volume of physiological saline and from each dilution 3 × 20 µL drop-plated on nutrient agar. Plates were incubated overnight at 30 °C after which colony forming units were counted. The experiment was repeated at least three times for each surface type and time point. Statistical analysis of test results were carried out in GraphPad Prism 7.04 software using one-way ANOVA analysis with Tukey's multiple comparisons test at 0.05 significance level. Due to highly variable and inconsistent results of lattice-embedded samples, these were excluded from statistical analysis at the 4 h time point.

3. Results

3.1. Structural Study and Mechanical Properties of the Composite Materials and Ceramics

The appearance of TiO_2 and TiO_2–2.5% Ag ceramic surfaces, and TiO_2–2.5% Ag in Ti6Al4V composite materials is shown in Figure 4. The white color of TiO_2 ceramic changed to gray when 2.5% Ag was added. However, lighter areas likely with lower Ag content can be seen in TiO_2–2.5% Ag ceramic (Figure 4). Under SEM (Figure 5) adequate interphase cohesion with some porosity between ceramic and Ti6Al4V lattice rods was observed. A SEM image (Figure 5a) and digital photograph (Figure 2) show that the 1 mm lattice structure was less bent or distorted subjected to SPS conditions. Critical areas were the interphases between the lattice and ceramic. Although water tightness of the hybrids was achieved, still some porosity at the interphase remained (Figure 5b).

Figure 4. Digital photograph of (**a**) ZrO$_2$, (**b**) pure TiO$_2$ anatase, (**c**) TiO$_2$–2.5% Ag, (**d**) composite structure with TiO$_2$–2.5% Ag and Ti6Al4V lattice after sintering (the lattice unit cell size is 1 mm and diameter of samples are 20 mm).

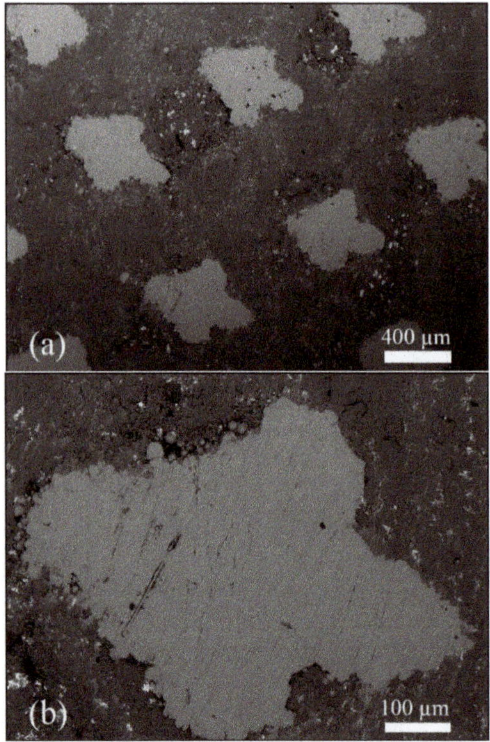

Figure 5. SEM micrograph of TiO$_2$–2.5% Ag and Ti6Al4V lattice structure (1 mm unit cell size) taken at ×50 magnification (**a**) and ×200 magnification (**b**).

The EDS elemental mapping results (Figure 6) showed presence of seven elements, namely, Ti 56.45%, O 39.36%, Ag 2.53%, V 0.56%, Al 0.48%, Cl 0.42%, and P 0.20%. The EDS spectrum illustrated acceptable distribution of TiO$_2$ and Ag in composition. Also, rounded Ti6Al4V rods cross-section validated the resistance of 1 mm cell size printed lattice under compression (Figures 2 and 6). To reveal the damage tolerance characteristics of ceramic–metal composite materials compared to pure ceramics, compression tests were performed (Figure 7). When plain titania ceramic showed brittle fracture, then lattice composite specimens did not catastrophically fail until 25% of deformation. For samples with larger volume fractions of metal phase (1 and 1.5 mm unit cell sizes) in addition to the absence of critical failure, ultimate strength of the composites was significantly higher.

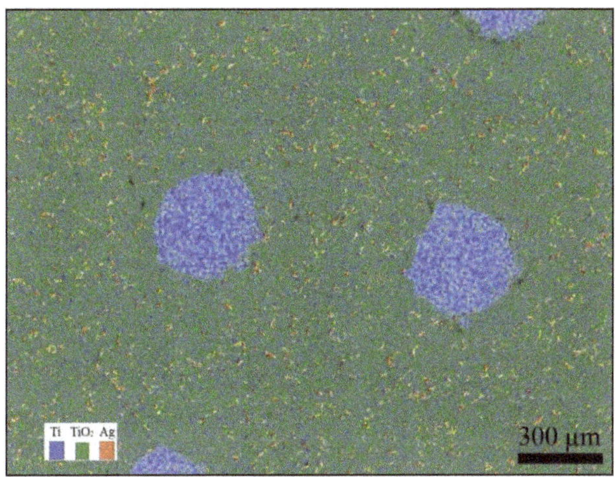

Figure 6. Energy dispersive spectroscopy (EDS) color mapping of TiO_2–2.5% Ag embedded 1 mm cell size Ti6Al4V lattice structure.

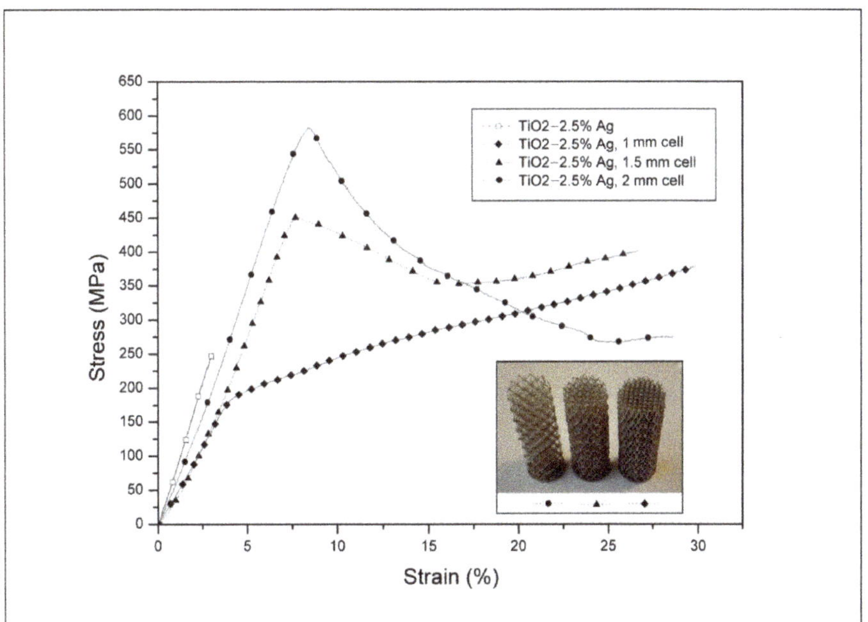

Figure 7. Compressive test results for TiO_2–Ag without and with different unit cell sizes of lattice structure. Height of lattices were 25 mm and diameter was 10 mm before SPS.

3.2. Antibacterial Activity of the Surfaces

Antibacterial activity of the composite and ceramic materials towards *E. coli* MG1655 was evaluated after 30 min and 4 h exposure (Figure 8) while using ceramic zirconia surface as a negative control. Among the tested materials, the highest antibacterial effect in dark conditions was observed for TiO_2–2.5% Ag ceramics where >3 logs reduction in *E. coli* viability was observed already within 30 min compared to control ($p < 0.01$) and no viable bacteria detected at the detection limit of about 450 colony forming units (CFU) per surface. As ceramic TiO_2 without added silver had no antibacterial effect in dark conditions ($p > 0.05$), we suggest that the effect seen for Ag-supplemented TiO_2 ceramics was due to Ag ions released from the material. Results for lattice-embedded TiO_2–2.5% Ag surfaces were not statistically different from control after 30 min ($p > 0.05$) and had too high variability to compare results with full ceramic materials after the 4 h time point. The fact that significantly less antibacterial effect was seen for lattice-embedded samples suggests that the release of Ag from those materials was much lower than from ceramic samples.

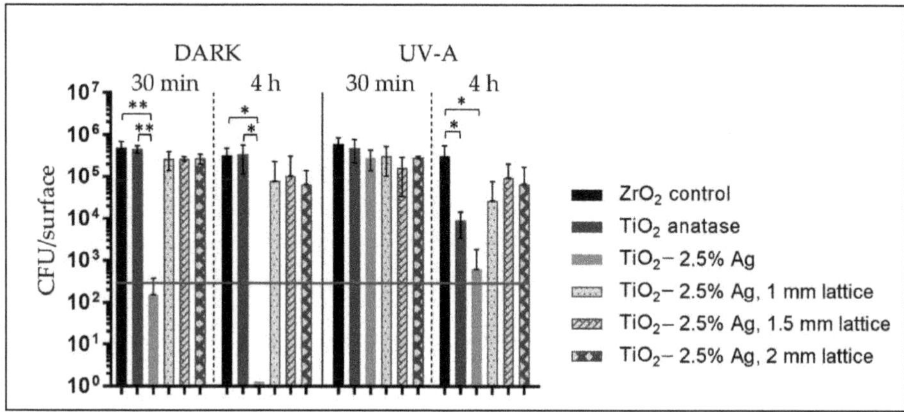

Figure 8. Viability of *Escherichia coli* MG1655 on ceramic and composite hybrid surfaces after 30 min and 4 h exposure in the dark and UV-A-illumination. Columns represent recovered viable bacteria as colony forming units (CFU). Mean and standard deviation of at least three independent values is shown on a logarithmic scale and only statistically significant differences ($p < 0.05$) marked on the graph (* $p < 0.05$ and ** $p < 0.01$). Only the upper error bar is shown for samples with >100% SD. Lattice samples are excluded from statistical analysis at 4 h time points due to very high variability. Limit of detection at 458 CFU/surface marked in red.

Due to the photocatalytic nature of TiO_2, the samples were assumed to exhibit UV-induced antibacterial effects. Indeed, 4 h exposure of bacteria to the ceramic TiO_2 surface under UV-A decreased bacterial viability by 1.6 logs ($p < 0.05$) compared to the control surface. The efficacy of Ag-supplemented ceramic TiO_2 surface under UV-A was higher than that of ceramic TiO_2 surface but significantly lower than the efficacy of TiO_2–2.5% Ag surface in dark conditions. This is likely because UV exposure has been shown to significantly decrease Ag solubility and subsequently, antibacterial activity as has been previously shown for photo-inducible Ag complemented ZnO surfaces [14].

Metal reinforced TiO_2–2.5% Ag ceramics showed significantly lower antibacterial effect than what was seen for ceramic surfaces. After 30 min under UV-A, TiO_2–2.5% Ag composite surfaces did not exhibit significant antibacterial effects compared with ZrO$_2$ control. After 4 h UV-A exposure, the composite surfaces yielded results that had too high variability to statistically compare them with control surfaces or full ceramic materials. However, according to the general picture, the composite surfaces exhibited slight antibacterial effect as compared to ZrO$_2$ control. These results showed that while including titanium lattice to TiO$_2$–2.5% Ag ceramic material increased the damage tolerance of the material, it significantly decreased the antibacterial effect of the ceramic material.

4. Discussion

Bonding of brittle ceramics to structural metals in assemblies has often remained a challenge, requiring designing for bolting or specific soldering alloys. The results described in this work represent a new approach for bonding, using additively manufactured lattice structure in the interphase of metal and ceramic. The composite structure where TiO$_2$–2.5% Ag was bonded with titanium showed not only increased damage tolerance but also increased ultimate compressive strength when compared to unreinforced ceramics. To explain the increased strength and damage tolerance, ceramic and metal–ceramic samples were subjected to a compressive test (Figure 7). The results showed large fractured pieces in the case of ceramics (Figure 9a), whereas ceramics in the composite sample was fractured into sub-micrometric particles (Figure 9b). For highly brittle material such as TiO$_2$–Ag, a larger content of energy is absorbed in crack initiation for hybrid composites.

Figure 9. Appearance of (**a**) TiO$_2$, (**b**) composite structure with TiO$_2$–Ti6Al4V hybrid after compressive testing (sample diameter was 10 mm and unit cell size was 1 mm).

The proposed failure mechanism of composite hybrids is visualized in Figure 10. Three modes of deformation could be distinguished. During the first mode, at applied strain up to 2–3 percent, energy was absorbed in rearrangement and elastic deformation of the metallic lattice. The elastic modulus of the composite hybrid (≈50–80 GPa) was significantly lower than the elastic modulus of both separated constituents, Ti6Al4V (≈110–120 GPa) and TiO$_2$ (≈230–280 GPa). In the middle region of the elastic part of the compressive loading curve a cracking sound was observed. This was due to the removal of fractures of ceramic elements exposed to the surface (schematically shown in Figure 10, Mode 2). Until the yielding point, surface exposed ceramic elements were gradually removed, and this remained as the main deformation mechanism during plastic deformation of the hybrid, during Mode 3. Additionally, the interior ceramic elements were fractured as shown in Figure 10, Mode 3. The opposing force from metallic lattice (indicated by black arrows in Figure 10) induced back-pressure and fractured ceramic pieces got embedded into a ductile metal lattice. This interaction increased the damage tolerance of the composite hybrid under compressive loading.

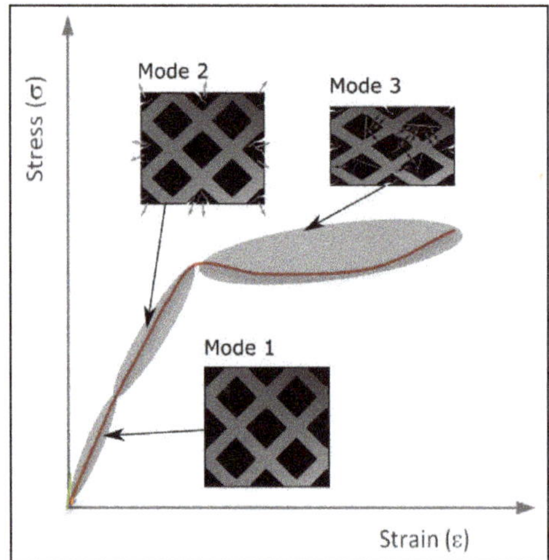

Figure 10. Schematic showing fracture mechanisms of metal–ceramic hybrids under compressive loading. Mode 1: rearrangement and elastic deformation of metal lattice; Mode 2: fracturing of ceramic surface elements; Mode 3: fracturing of the interior ceramic elements and embedding of these in ductile metal lattice.

Consequently, three modes of failure of the hybrid composite were differentiated during compressive loading. The increased damage tolerance and compressive strength were attributed to higher input energy needed to fracture the ceramics and interaction between fractured ceramic pieces and ductile metal lattice inside the material. As it was seen on the hybrid specimens after compressive testing (Figure 9b), the ceramic material was removed preferentially at the perimeter of the cylindrical sample. Further strengthening could be achieved if hybrid composite would be surrounded by an additional metallic layer so that ceramic would not be exposed on the outer surface.

Metal 3D printing enables the production of lattice structure objects with different shapes and internal mesostructures. The composite, therefore, can be designed according to existing mechanical loads. Functionally grading in different directions, and integrating solid printed metals with ceramic–metal hybrids could be realized. Furthermore, a metal lattice could be designed so that it acts as a heating element when an electric current is directed through the material. The heating would further enhance the antimicrobial effect of the ceramic.

The compressive test result of TiO_2 ceramic and Ti6Al4V–TiO_2 lattice composite was captured (Figure 10) and showed the benefits of lattice structure perfectly. Finite element analysis prepares an estimation of metal–ceramic composite materials strength produced by combining SLM-SPS technique subjected impact, abrasion or compression loading [15–17]. Numerical simulation illustrated that fracture will occur at around 500 MPa for pure ceramic, while buckling for metallic lattice will start from 1200 MPa (Figure 11). Damage tolerance of Ti6Al4V–TiO_2 composite depends on the densification of ceramic (SLM parameters) and metal–ceramic bonding phase (SPS parameters).

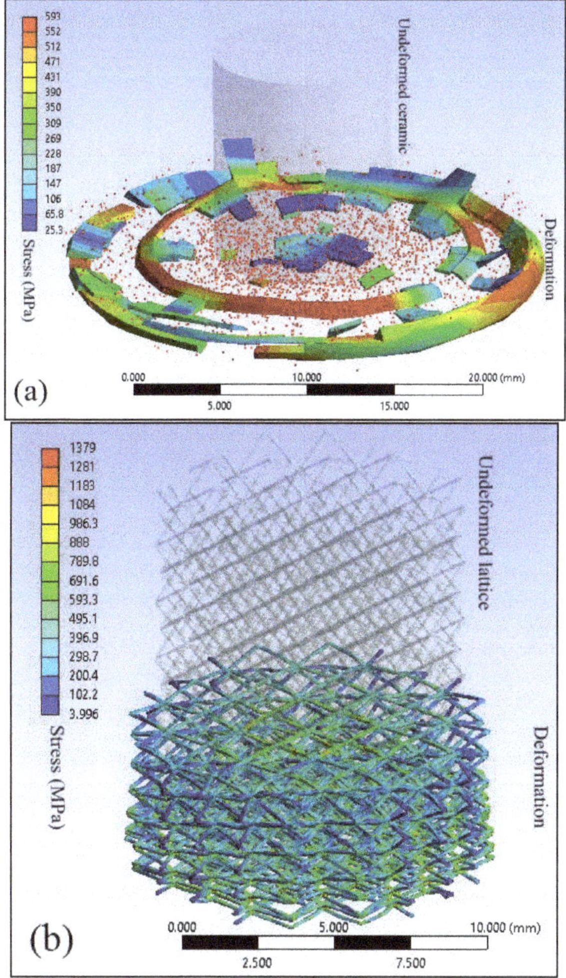

Figure 11. Compressive strength modelling of (**a**) TiO_2 ceramic, (**b**) Ti6Al4V lattice structure. Simulation conditions and dimensions are identical for the ceramic and lattice structures.

The present work used SPS as a consolidation method, which can produce only simple, cylindrical shapes. Using other hot consolidation methods as hot isostatic pressing or hot forging, 3D shaped hybrid composites could also be manufactured. The ability to produce composites with complex shape could provide new solutions for a number of applications in the field of metal–ceramic hybrids.

This study was unique as there are no prior studies reporting on antibacterial effects of hybrid composite metal–ceramic materials. However, studies on antibacterial effects of Ag-containing ceramic materials have been previously published [18]. In general, those studies have shown the relationship between the amount of silver in the ceramic surface and antibacterial activity [19] and the importance of segregation and agglomeration of silver on the surface for improving antibacterial efficacy [20–22]. Similar observation was also done in this study but only in dark conditions. Under UV-A, the effect of added Ag to ceramics had significantly smaller effect than in dark conditions. We suggest that this was due to Ag ions which drive the antibacterial effect of Ag–ceramic surfaces [23–25] being reduced back to elemental Ag onto the surfaces [14]. Compared with TiO_2–Ag ceramic material, the antimicrobial effect of hybrid surfaces was drastically reduced (Figure 8). This change could not be only explained by reduced area of antimicrobial TiO_2–Ag surface in composite material as titanium metal lattice occupied 15% to 25% of the surface, depending on the sample. In almost all cases, both in dark and under ultraviolet exposure the antimicrobial effect of composite surfaces was orders of magnitude lower when compared to the fully ceramic surface. The reason for this was not clear, but it could be assumed that there was a combination of direct surface contact and soluble silver toxicity in effect, both dependent on silver exposure at the material surface. These results indicated that if the hybrid composite needed to be exposed on the surface, the metal composition would also need to be chemically modified for an antimicrobial effect. Otherwise, we suggest that in order to preserve the antibacterial activity of the composite material and reduce variability in antibacterial activity results, a thin layer of pure ceramic material should be added to the surface of the composite.

5. Conclusions

An approach to produce titanium/silver-doped titania composites was introduced, by combining of SLM and SPS techniques. The metallic lattice structures were 3D printed, embedded with TiO_2–Ag ceramic powder and consolidated by SPS. Compression strength and damage tolerance of the composites were shown to increase significantly when compared to TiO_2–Ag ceramics. No collapsing of the composites was seen at up to 25% of deformation in the compressive test. Adding metallic lattice to ceramic silver-doped titania material however decreased the antibacterial effect compared with ceramics only, significantly. Thus, we suggest that the composition of metal that is used to produce the lattice should be chemically modified or a thin layer of pure ceramic material should be added to the surface of the composite.

Author Contributions: For research articles with several authors, a short paragraph specifying their individual contributions must be provided. The following statements should be used "conceptualization, R.R. and M.R.; methodology, L.K. and A.I.; software, R.R.; validation, M.R., A.I. and L.K.; formal analysis, R.R.; investigation, R.R. and M.R.; resources, L.K. and A.I.; data curation, M.R.; writing—original draft preparation, R.R.; writing—review and editing, R.R., M.R., A.I. and L.K.; visualization, A.I.; supervision, L.K.; project administration, A.I.; funding acquisition, L.K.".

Funding: This research was funded by the Estonian Ministry of Education and Research (IUT 19-29; PUT 748; IUT 23-5; base funding provided to R&D institutions B56 and SS427; M-ERA.NET DURACER project ETAG18012); The European Regional Fund, project number 2014-2020.4.01.16-0183 (Smart Industry Centre) and ERDF project TK134.

Acknowledgments: The authors would like to thank Mart Viljus for the help with EDS mapping.

Conflicts of Interest: The authors declare that there are no conflicts of interest.

References

1. He, C.; Yu, Y.; Hu, X.; Larbot, A. Influence of silver doping on the photocatalytic activity of titania film. *Appl. Surf. Sci.* **2002**, *200*, 239–247. [CrossRef]
2. Sung-Suh, H.M.; Choi, J.R.; Hah, H.J.; Koo, S.M.; Bae, Y.C. Comparison of Ag deposition effects on the photocatalytic activity of nanoparticulate TiO_2 under visible and UV light irradiation. *J. Photochem. Photobiol. Chem.* **2004**, *163*, 37–44. [CrossRef]
3. Zhao, L.; Wang, H.; Huo, K.; Cui, L.; Zhang, W.; Ni, H.; Zhang, Y.; Wu, Z.; Chu, P.K. Antibacterial nano-structured titania coating incorporated with silver nanoparticles. *Biomaterials* **2011**, *32*, 5706–5716. [CrossRef] [PubMed]
4. Oja Acik, I.; Junolainen, A.; Mikli, V.; Danilson, M.; Krunks, M. Growth of ultra-thin TiO_2 films by spray pyrolysis on different substrates. *Appl. Surf. Sci.* **2009**, *256*, 1391–1394. [CrossRef]
5. Kim, H.C.; Park, H.K.; Shon, I.J.; Ko, I.Y. Fabrication of ultra-fine TiO_2 Ceramics by a high-frequency induction heated sintering method. *J. Ceram. Process. Res.* **2006**, *7*, 327.
6. Matsuno, H.; Yokoyama, A.; Watari, F.; Uo, M.; Kawasaki, T. Biocompatibility and osteogenesis of refractory metal implants, titanium, hafnium, niobium, tantalum and rhenium. *Biomaterials* **2001**, *22*, 1253–1262. [CrossRef]

7. Asa'ad, F.; Pagni, G.; Pilipchuk, S.P.; Giannì, A.B.; Giannobile, W.V.; Rasperini, G. 3D-Printed Scaffolds and Biomaterials: Review of Alveolar Bone Augmentation and Periodontal Regeneration Applications. *Int. J. Dent.* **2016**, *2016*, 1239842. [CrossRef]

8. Vandenbroucke, B.; Kruth, J.P. Selective laser melting of biocompatible metals for rapid manufacturing of medical parts. *Rapid Prototyp. J.* **2007**, *13*, 196–203. [CrossRef]

9. Könönen, M.; Kivilahti, J. Concise review biomaterials & bioengineering: Fusing of dental ceramics to titanium. *J. Dent. Res.* **2001**, *80*, 848–854.

10. Zhang, C.; Chaudhary, U.; Lahiri, D.; Godavarty, A.; Agarwal, A. Photocatalytic activity of spark plasma sintered TiO_2-graphene nanoplatelet composite. *Scr. Mater.* **2013**, *68*, 719–722. [CrossRef]

11. Lee, Y.; Lee, J.H.; Hong, S.H.; Kim, D.Y. Preparation of nanostructured TiO_2 ceramics by spark plasma sintering. *Mater. Res. Bull.* **2003**, *38*, 925–930. [CrossRef]

12. Noh, J.H.; Jung, H.S.; Lee, J.K.; Kim, J.R.; Hong, K.S. Microwave dielectric properties of nanocrystalline TiO_2 prepared using spark plasma sintering. *J. Eur. Ceram. Soc.* **2007**, *2*, 2937–2940.

13. Bernasconi, R.; Carrara, E.; Hoop, M.; Mushtaq, F.; Chen, X.; Nelson, B.J.; Pané, S.; Credi, C.; Levi, M.; Magagnin, L. Magnetically navigable 3D printed multifunctional microdevices for environmental applications. *Addit. Manuf.* **2019**, *28*, 127–135. [CrossRef]

14. Visnapuu, M.; Rosenberg, M.; Truska, E.; Nommiste, E.; Sutka, A.; Kahru, A.; Rahn, M.; Vija, H.; Orupold, K.; Kisand, V.; et al. UVA-induced antimicrobial activity of ZnO/Ag nanocomposite covered surfaces. *Colloids Surf. B Biointerfaces* **2018**, *169*, 222–232. [PubMed]

15. Rahmani, R.; Antonov, M.; Kamboj, N. Modelling of impact-abrasive wear of ceramic, metallic, and composite materials. *Proc. Est. Acad. Sci.* **2019**, *68*, 191–197.

16. Rahmani, R.; Antonov, M.; Kollo, L. Wear Resistance of (Diamond-Ni)-Ti6Al4V Gradient Materials Prepared by Combined Selective Laser Melting and Spark Plasma Sintering Techniques. *Adv. Tribol.* **2019**, *2019*, 5415897. [CrossRef]

17. Rahmani, R.; Antonov, M.; Kollo, L. Selective Laser Melting of Diamond-Containing or Postnitrided Materials Intended for Impact-Abrasive Conditions: Experimental and Analytical Study. *Adv. Mater. Sci. Eng.* **2019**, *2019*, 4210762. [CrossRef]

18. Velasco, S.C.; Cavaleiro, A.; Carvalho, S. Functional properties of ceramic-Ag nanocomposite coatings produced by magnetron sputtering. *Prog. Mater. Sci.* **2016**, *84*, 158–191. [CrossRef]

19. Hsieh, J.H.; Tseng, C.C.; Chang, Y.K.; Chang, S.Y.; Wu, W. Antibacterial behavior of TaN-Ag nanocomposite thin films with and without annealing. *Surf. Coat. Technol.* **2008**, *202*, 5586–5589.

20. Wickens, D.J.; West, G.; Kelly, P.J.; Verran, J.; Lynch, S.; Whitehead, K.A. Antimicrobial activity of nanocomposite zirconium nitride/silver coatings to combat external bone fixation pin infections. *Int. J. Artif. Organs.* **2012**, *35*, 817–825.

21. Hsieh, J.H.; Chang, C.C.; Li, C.; Liu, S.J.; Chang, Y.K. Effects of Ag contents on antibacterial behaviors of TaON-Ag nanocomposite thin films. *Surf. Coat. Technol.* **2010**, *205*, 337–340.

22. Huang, H.L.; Chang, Y.Y.; Lai, M.C.; Lin, C.R.; Lai, C.H.; Shieh, T.M. Antibacterial TaN-Ag coatings on titanium dental implants. *Surf. Coat. Technol.* **2010**, *205*, 1636–1641. [CrossRef]

23. Jamuna-Thevi, K.; Bakar, S.A.; Ibrahim, S.; Shahab, N.; Toff, M.R.M. Quantification of silver ion release, in vitro cytotoxicity and antibacterial properties of nanostuctured Ag doped TiO2 coatings on stainless steel deposited by RF magnetron sputtering. *Vacuum* **2011**, *86*, 235–241. [CrossRef]

24. Song, D.H.; Uhm, S.H.; Kim, S.E.; Kwon, J.S.; Han, J.G.; Kim, K.N. Synthesis of titanium oxide thin films containing antibacterial silver nanoparticles by a reactive magnetron co-sputtering system for application in biomedical implants. *Mater. Res. Bull.* **2012**, *47*, 2994–2998.

25. Ferraris, M.; Miola, S.; Ferraris, S.; Gautier, G.; Maina, G. Chemical, mechanical, and antibacterial properties of silver nanocluster-silica composite coatings obtained by sputtering. *Adv. Eng. Mater.* **2010**, *12*, 276–282.

Article

Controlling Nitrogen Dose Amount in Atmospheric-Pressure Plasma Jet Nitriding

Ryuta Ichiki [1,*]**, Masayuki Kono** [1]**, Yuka Kanbara** [1]**, Takeru Okada** [2]**, Tatsuro Onomoto** [3]**,
Kosuke Tachibana** [1]**, Takashi Furuki** [1] **and Seiji Kanazawa** [1]

[1]   Division of Electrical and Electronic Engineering, Oita University, Oita 870-1192, Japan;
v1652026@oita-u.ac.jp (M.K.); v19e2007@oita-u.ac.jp (Y.K.); tachibana-kosuke@oita-u.ac.jp (K.T.);
furuki-takashi@oita-u.ac.jp (T.F.); skana@oita-u.ac.jp (S.K.)

[2]   Department of Electronic Engineering, Tohoku University, Sendai 980-8579, Japan;
takeru.okada@tohoku.ac.jp

[3]   Material Technology Division, Fukuoka Industrial Technology Center, Kitakyushu 807-0831, Japan;
onomoto@fitc.pref.fukuoka.jp

*   Correspondence: ryu-ichiki@oita-u.ac.jp; Tel.: +81-97-554-7826

Received: 25 April 2019; Accepted: 24 June 2019; Published: 25 June 2019

Abstract: A unique nitriding technique with the use of an atmospheric-pressure pulsed-arc plasma jet has been developed to offer a non-vacuum, easy-to-operate process of nitrogen doping to metal surfaces. This technique, however, suffered from a problem of excess nitrogen supply due to the high pressure results in undesirable formation of voids and iron nitrides in the treated metal surface. To overcome this problem, we have first established a method to control the nitrogen dose amount supplied to the steel surface in the relevant nitriding technique. When the hydrogen fraction in the operating gas of nitrogen/hydrogen gas mixture increased from 1% up to 5%, the nitrogen density of the treated steel surface drastically decreased. As a result, the formation of voids were suppressed successfully. The controllability of the nitrogen dose amount is likely attributable to the density of NH radicals existing in the plume of the pulsed-arc plasma jet.

Keywords: plasma nitriding; atmospheric-pressure plasma; nitrogen dose amount; hydrogen fraction; void

---

1. Research Background

Plasma nitriding is a surface technology to dope nitrogen atoms into metal surfaces via plasma chemical reactions to improve wear resistance, and fatigue strength, etc., of materials [1–16]. Plasma nitriding is now one of the essential surface treatments used in industry, especially in the automobile industry and die/mold fabrication. Conventional plasma nitriding uses low-pressure DC (or pulsed DC) plasmas in the abnormal glow discharge mode, where the batch process with a large vacuum furnace meets the purpose of mass production. In addition, a number of low-pressure plasma modes have recently been applied to nitriding treatment, e.g., active screen plasmas [4–6], electron cyclotron resonance plasmas [2,7], and radio-frequency plasmas [8], etc.

As another technological seed, nitriding methods using atmospheric-pressure plasmas have been developed, where the disuse of vacuum equipment makes the process much quicker and easier-to-operate. Two types of atmospheric-pressure plasmas are utilized to nitriding, namely the pulsed-arc (PA) plasma jet [17–22] and the dielectric barrier discharge (DBD) [23,24]. The PA plasma-jet nitriding has proved to be available to die steel [17,18,22], austenitic stainless steel [20], and titanium alloy [19,21], where the jet plume is sprayed onto the sample surface to thermally diffuse nitrogen atoms into it. Note that the nitrogen/hydrogen gas mixture is used as the operating gas. The plasma-jet nitriding will offer a drastically economical method to us compared with conventional plasma nitriding, especially when high-mix low-volume production is targeted.

The plasma-jet nitriding, however, is a relatively new, still developing technology. Thus, its controllability and reliability has to be improved further for practical application. For example, we had no methods to control the nitrogen dose amount from the jet plume to the metal surface, while such a method has been completed for conventional nitriding in which the nitrogen dose is well-controlled by adjusting the nitriding potential [25]. Due to the lack of dose controllability, the plasma-jet nitriding suffers from a problem of excess nitrogen supply due to the high pressure results in undesirable formation of voids and iron nitrides (the compound layer) attributed to nitrogen gas precipitation in the treated metal surface.

In this paper, a newly developed method is detailed to control nitrogen dose amount in plasma-jet nitriding to overcome the problem of excess nitrogen supply. A brief introduction of the method is as follows. The operating gas to generate the plasma jet is a nitrogen/hydrogen gas mixture. The optical emission spectroscopy proved that NH radical emission is dominant from the jet plume. In general, the NH emission intensity tends to decrease with increasing hydrogen fraction in the operating gas, f_{H2}. If NH is the key radical for plasma-jet nitriding and if the decreasing tendency of NH emission with f_{H2} indicates decreasing NH density in the jet plume, we could decrease the nitrogen dose amount to metal surface by increasing f_{H2}. Following this assumption, we addressed to control the nitrogen dose amount by changing f_{H2} in this study.

Prior to explaining our research, let us summarize here key species in various plasma nitriding techniques. As for low-pressure plasma nitriding methods, a comprehensive and systematic understanding of key species is not present. For example, several papers suggest the importance of ion species such as N^{2+} [2], N^+ [9], and NH_x^+ [10]. On the other hand, Matsumoto et al. proposed that neutral species govern the rate-limiting step [11]. For the radical nitriding, one of the low-pressure plasma nitriding methods using NH_3 and H_2 gas, NH radicals are considered to play a key role [12,13]. Besides, some other papers mention the importance of NH radicals for plasma nitriding [8,14]. Moreover, in gas nitriding, NH_3 dissociates on an iron surface to NH_2, NH_2 dissociates to NH, and NH dissociates to N and H, in order, indicating that the presence of NH is essential for gas nitriding [25]. As described above, a number of studies regard NH radicals as species effective to nitriding. Thus, our expectation that NH is key for plasma-jet nitriding is not peculiar.

2. Experimental Procedure

Experiments were performed with the PA plasma jet system shown in Figure 1a. The jet nozzle was composed of coaxial cylindrical electrodes. The grounded external electrode measured 35 mm in inner diameter. The discharge gap was approximately 20 mm. The nitrogen/hydrogen gas mixture was introduced into the nozzle at the total flow rate of 20 slm, where the hydrogen fraction, and the ratio of hydrogen flow rate to total flow rate, was f_{H2}. The low-frequency voltage pulse (5 kV in height and 21 kHz in repetition) was applied to the inner electrode, resulting in the maximum of the discharge current of ca. 1.2 A. The afterglow of the generated PA plasma was spewed out from the orifice and was 4 mm in diameter, forming the jet plume containing NH radicals.

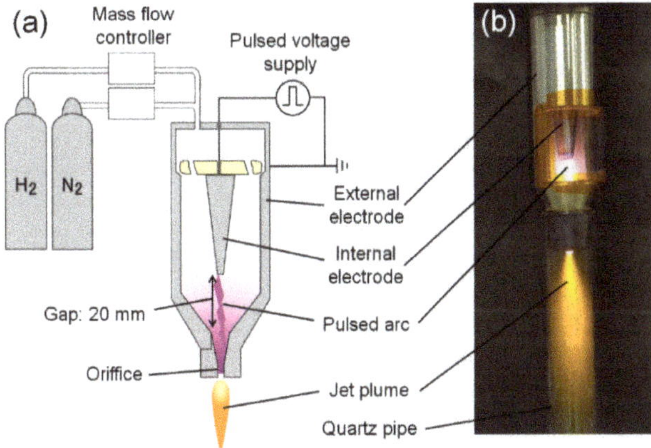

Figure 1. Pulsed-arc plasma jet. (**a**) Schematic of jet nozzle. (**b**) Photograph of plasma jet.

For the spectroscopic experiment, the jet nozzle was fitted with a quartz pipe (30 mm in diameter and 500 mm in length) at the tip as shown in Figure 1b to observe the jet plume generating in the operating gas. The optical emission of NH ($A^3\Pi$–$X^3\Sigma^-$) of 336 nm was detected with a spectrometer (Shamrock SR-500i, Andor, Belfast, UK). We collected the light emitted from the jet plume at the distance of 10 mm from the nozzle tip.

For the nitriding experiment, the jet nozzle was inserted into a cylindrically shaped cover made of quartz as shown in Figure 2 to purge residual oxygen from the treatment atmosphere. The height and diameter of the quartz cover was 85 and 124 mm, respectively. The gas was exhausted through the 1-mm gap between the jet nozzle and the quartz cover. The experimental system was put in a simple booth ($1.0 \times 1.2 \times 1.8$ (height) m^3) surrounded by a vinyl curtain to lead the exhaust gas to the gas-treatment equipment. Prior to generating the plasma jet, residual oxygen inside the cover was gas-purged by the operating gas introduced through the nozzle.

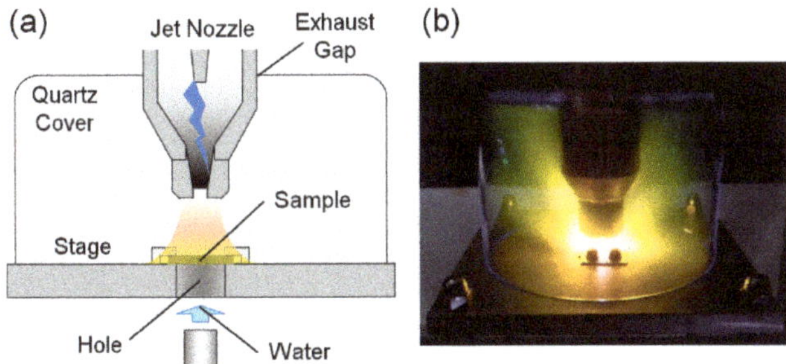

Figure 2. Experimental Setup with the quartz cover to purge residual oxygen. (**a**) Schematic. (**b**) Photograph of plasma-jet nitriding.

The steel to be treated was cold roll steel JIS SPCC. The composition was as follows: 0.02% C, 0.09% Mn, 0.017% P, 0.004% S, and the balance was Fe. The sample dimension was $25 \times 25 \times 1.2$ mm^3. The hardness of base material was ca. 150 HV. The surface was mirror finished with alumina powder (1 μm) and degreased in an ultrasonic acetone bath.

To make the effects of nitrogen dose amount as conspicuous as possible, the surface temperature was set into the range of 1000 to 1100 K during nitrogen doping and the doped sample was immediately quenched to invoke iron-nitrogen martensite transformation. Such nitro-quenching treatment was known to form voids, which can be readily observed with a microscope, in the surface when excess nitrogen was doped [26]. In addition, the formation of iron-nitrogen martensite indicated to us the answer as to whether or not a non-trivial amount of nitrogen had been doped even when the nitrogen dose amount was intently reduced to suppress the formation of voids. The surface temperature of ca. 1000–1100 K was maintained by the plasma-jet spraying itself, where the distance between the nozzle tip to the surface was set to 7 mm. The treatment temperature was measured by spraying the jet plume to a dummy sample with a thermocouple on the surface. The doping duration was 900 to 1800 s. The doped steel was quenched by water cooling, where tap water was poured onto the opposite surface through the hole bored in the center of the sample stage.

The doped nitrogen concentration in the steel surface was detected by an electron probe micro analyzer (EPMA, JXA-8200SP, JEOL, Tokyo, Japan). The void formation and the metallographic structure were observed with an optical microscope (VHX-5000, KEYENCE, Osaka, Japan) to a cross-section of doped steel surface. The sample surface was etched in nital solution (3%) for observing the metallographic structure. The formation of iron nitrides was detected by X-ray diffraction (XRD, SmartLab, Rigaku, Tokyo, Japan) using Co Kα radiation (λ = 0.179 nm). The hardness profile of the cross-section was measured with a Vickers microhardness tester (FM-300, FUTURE-TECH, Kawasaki, Japan), where the indenter load was 0.098 N and the loading time was 10 s.

3. Results and Discussions

3.1. NH Emission Intensity

Figure 3 shows the f_{H2} dependence of the emission intensity of NH radicals from the jet plume. The NH emission intensity increased with increasing f_{H2} up to 0.25%. On the contrary, the NH emission intensity turned to decrease with increasing f_{H2} over 0.25% up to 5%. The decreasing tendency of the emission intensity suggested the likely possibility that the density of NH radical existing in the jet plume decreased with increasing f_{H2} in the range more than 0.25%. Following this suggestion, we increased f_{H2} for the purpose of decreasing the nitrogen dose amount. Incidentally, the minimum f_{H2} was set to 1% in this study because by our previous work, f_{H2} was less than 1% proved to result in oxidization of the steel surface due to the lack of reduction performance against residual oxygen [18].

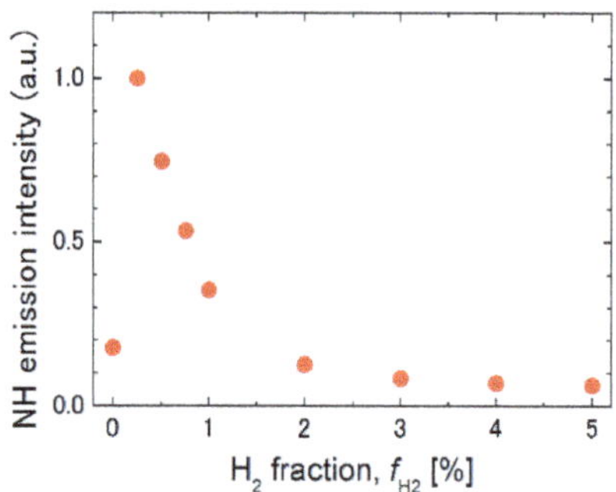

Figure 3. Emission intensity of NH radicals (336 nm) in jet plume vs H_2 fraction in the operating gas.

3.2. Nitrogen Density of Treated Steel Surface

Figure 4a,b shows the cross-sectional mapping and depth profile of nitrogen concentration, respectively, where the observation point was at the center of plasma-jet spraying. For f_{H2} of 1%, nitrogen was considerably condensed in the surface. In the vicinity of the outermost surface, the nitrogen concentration reached ca. 12 at% and monotonically decreased in the depth direction. The concentration gradient was a typical characteristic in such a diffusion treatment. On the other hand, the nitrogen concentration in the vicinity of the surface became obviously less for f_{H2} of 2.5%

even though the gradient tendency was analogous. The maximum concentration was merely 4 at% in this case. Moreover, further decreases in nitrogen concentration was seen for f_{H2} of 5%, where the maximum value was reduced down to 2 at%. In summary, the nitrogen concentration in the doped steel surface monotonically decreased with increasing f_{H2}, while it maintained the gradient tendency in the depth direction. This result indicates that the nitrogen dose amount was successfully controlled by changing f_{H2}.

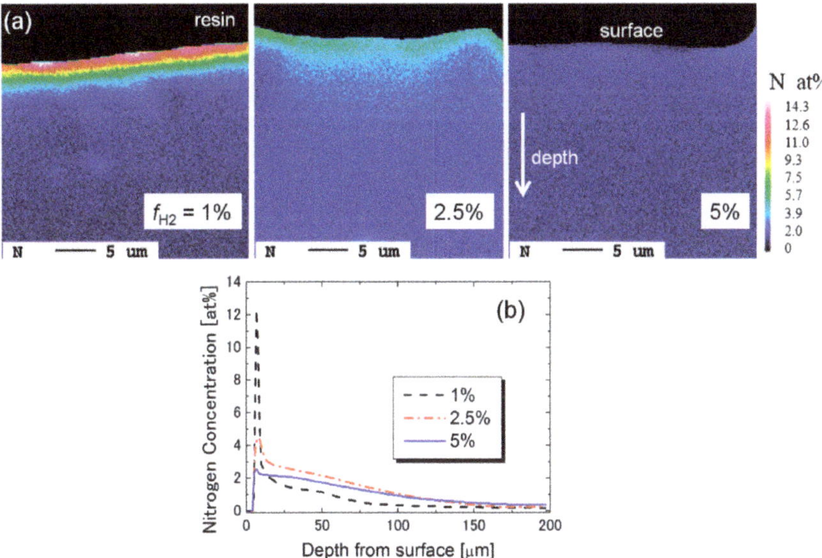

Figure 4. Distribution of nitrogen concentration for several f_{H2}. The doping duration was 1800 s. (**a**) Two-dimensional mapping of sample cross-section in the vicinity of treated surface. (**b**) One-dimensional depth profile.

3.3. Formation of Voids

Figure 5 shows cross-sectional micrographs of doped steels observed in the vicinity of the center of plasma-jet spraying. For f_{H2} of 1%, a number of black dots were seen, which corresponded to the voids due to excess nitrogen doping. The existence of such voids would have made the material surface extremely brittle. On the other hand, the number of voids tended to decrease with increasing f_{H2} and as a consequence, they become invisible in the optical microscopic scale for f_{H2} of 5%. This result indicates that increasing f_{H2} can suppress the void formation in the steel surface. From the tendency of nitrogen concentration described in Section 3.2, it follows that decreasing the nitrogen dose amount was the cause of the suppression of void formation.

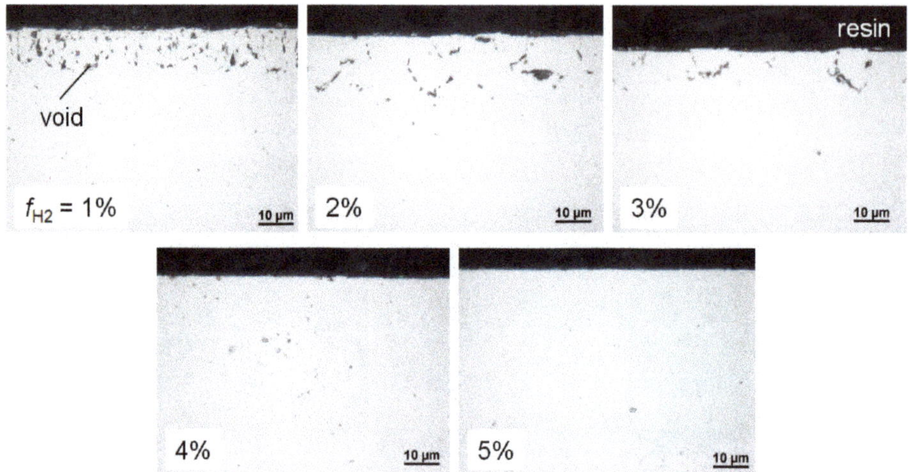

Figure 5. Micrographs of sample cross-section. The dots appearing in the vicinity of surface correspond to voids. The doping duration was 900 s.

3.4. Formation of Compound Layer

Figure 6 shows XRD spectra of the treated steel surface, where the observation point was the sample surface in the vicinity of the center of plasma-jet spraying. For f_{H2} of 1%, the treated surface obviously contained iron nitrides, namely Fe_4N (γ' phase) and $Fe_{2-3}N$ (ε phase). On the other hand, the spectral intensities of the iron nitrides suddenly decreased with increasing f_{H2} and as a consequence, they became less than the detection limit for f_{H2} over 2%. This result indicated that increasing f_{H2} reduced the formation of iron nitrides in the steel surface. From the tendency of nitrogen concentration described in Section 3.2, it follows that the formation of iron nitrides was suppressed owing to a decreasing nitrogen dose amount.

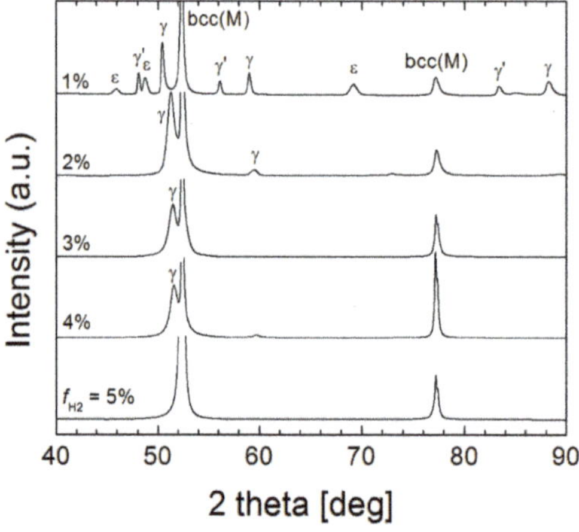

Figure 6. XRD spectra of treated steel surface in the vicinity of the center of plasma-jet spraying.

In addition, the XRD peaks of the retained austenite (γ phase) were clearly seen. The formation of the retained austenite was attributed to the austenitic transformation over the critical temperature A_1 and an excess solution of nitrogen. The formation of retained austenite can be regarded as another negative effect of excess nitrogen supply as well as the voids and iron nitrides. However, Figure 6 exhibits that the peak intensity of γ tended to decrease with increasing f_{H2}, indicating that the amount of retained austenite was reduced. Moreover, the γ peak shifts toward high theta with f_{H2}, that resulted from decreasing the lattice constant due to decreasing dissolved nitrogen concentration. From the relationship between the austenitic lattice constant a and the dissolved nitrogen concentration X_N in atomic percentage ($a/nm = 0.3564 + 0.00077X_N$) [26], we obtained the dependence of X_N on f_{H2} as shown in Figure 7. For $f_{H2} = 1\%$, $X_N = 9.6$ at%. Increasing f_{H2} monotonically decreased X_N and for $f_{H2} = 4\%$, X_N was reduced down to 0.26 at%. These characteristics of retained austenite are additional evidence for the controllability of nitrogen dose amount.

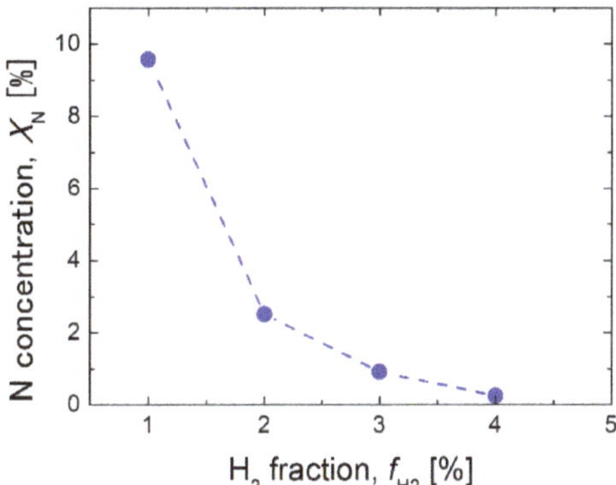

Figure 7. Nitrogen concentration in retained austenite calculated from the XRD spectral shift.

3.5. Hardness Profile

Figure 8 shows the two-dimensional hardness profiles of the cross-section of treated steels. The horizontal axis is the surface position of sample, the origin of which corresponds to the center of plasma-jet spraying. The vertical axis is the depth from surface. Here the micro-Vickers hardness was measured at intervals of 2 mm in the horizontal direction and 10 μm in the vertical direction. The hardness is displayed by gray scale.

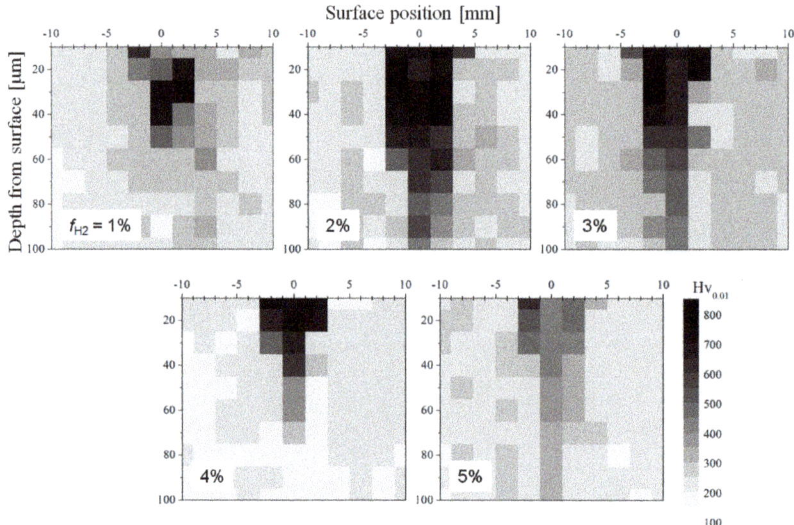

Figure 8. Two-dimensional hardness profiles of sample cross-section. The micro-Vickers hardness is displayed by gray scale. The doping duration was 900 s.

We see that for every f_{H2}, the hardness beneath the center of plasma-jet spraying was increased significantly beyond the original hardness. Here the area of ca. 5 mm in diameter was locally hardened. The local hardening was most likely due to the limited heating ability of the plasma jet only to a narrow area, not due to local nitrogen supply because of the following fact. It has already been proved that nitrogen can be supplied to the circular area as large as 20 mm in diameter by identical plasma-jet spraying, where the steel sample was heated up to ca. 800 K with the assistance of an external heater [18]. The highest hardness was 815, 755, 815, 822, and 606 HV for f_{H2} = 1%, 2%, 3%, 4%, and 5%, respectively. Such hardness cannot be obtained without nitrogen doping because the original carbon content was too low in the sample to invoke iron-carbon martensite transformation. Although the hardness was relatively low when f_{H2} = 5%, the drastic increase in hardness proved that the nitrogen dose amount was still appropriate.

Figure 9 shows the depth profile of hardness averaged within the range of ±2.5 mm of the surface position. The error bar corresponds to the standard deviation of each of the three data sets. We can see the typical trend of hardness gradient for every f_{H2}. Note that for f_{H2} = 1%, obvious softening occurs in the outermost surface within the depth profile from 10 to 30 μm. (This is clearly seen also in Figure 8.) This softening was possibly due to the considerable amount of retained austenite. For f_{H2} = 2%, the softening effect became much weaker and for more f_{H2} it was not seen any more. The dependence of the softening effect on f_{H2} is consistent with the peak intensity of γ shown in Figure 6. The outermost hardness of f_{H2} = 3% and 4% reached the largest value of ca. 800 HV. For f_{H2} = 5% it became lesser, likely due to a lower nitrogen dose amount.

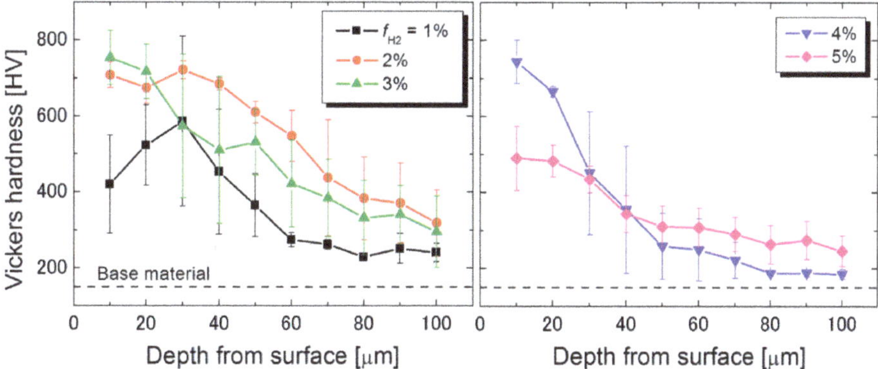

Figure 9. Depth profile of hardness of sample cross-section. The error bars correspond to the standard deviation.

3.6. Metallographic Structure

Figure 10 shows the metallographic micrographs of sample cross-section in the vicinity of the surface. Here the martensite layer and the compound layer are denoted by m and c, respectively. For f_{H2} = 1%, the compound layer was clearly seen as a discontinuous thin layer. However, no discontinuous layer appeared for more f_{H2}, being consistent with the behavior of γ' and ε peaks shown in Figure 6.

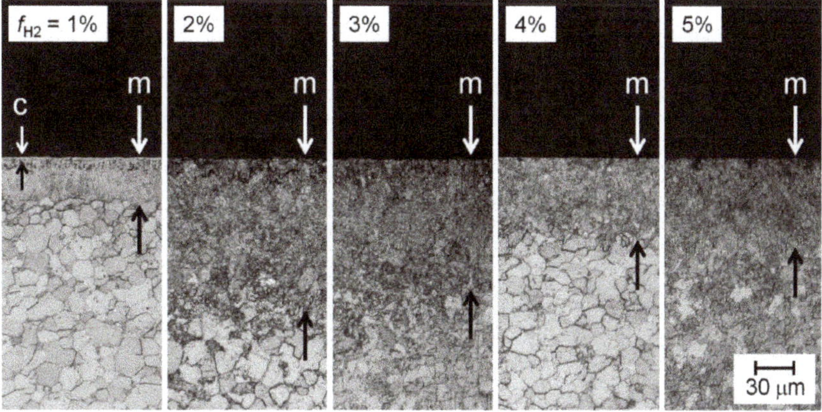

Figure 10. Metallographic micrograph of sample cross-section. The arrow pairs denoted by m and c specify the vertical range of the martensite layer and the compound layer, respectively.

The thickness of the martensite layer depended on f_{H2}. That is, the thickness increased with increasing f_{H2} up to 2%, and turned to a decrease from 2% to 4%, and then kept constant for more f_{H2}. We consider that the thickness change was caused by a shift of the surface temperature. The temperature of jet plume tended to increase with increasing f_{H2} owing to the high thermal conductivity of hydrogen, transporting the thermal energy from the pulsed arc to the jet plume. The sample temperature (measured with a dummy sample) was ca. 1000 K for f_{H2} = 1% but increased up to ca. 1100 K for more f_{H2}. From this fact, it follows that the increase in the thickness from f_{H2} = 1% to 2% was caused by the enhanced thermal diffusion and the subsequent decrease was caused by the decrease in nitrogen dose amount. The change in the layer thickness can be seen also in Figures 8 and 9.

4. Conclusions

To overcome the problem of excess nitrogen supply in the PA plasma-jet nitriding, a controlling method of nitrogen dose amount was proposed on the basis of NH radical emission and was addressed to experimentally performed. Consequently, we have demonstrated that the nitrogen dose amount to the steel surface can be controlled by changing the hydrogen fraction in the operating gas. As a result of nitrogen dose control, undesirable formation of voids and iron nitrides were successfully suppressed, while a nitrogen dose enough to invoke martensite transformation was simultaneously achieved.

This achievement means that we have first obtained the technique to control "the nitriding potential" even in the new plasma-jet nitriding. We believe that the upgraded controllability presented here was of great help for future practical applications of the plasma-jet nitriding, especially for applications to high-mix low-volume production of mechanical and medical fabrications. Note that although the treatable area in this study seems too small for practical use, we can practically treat the circular area of at least 20 mm in diameter for the ordinary nitriding temperature at ca. 800 K [18].

Author Contributions: Conceptualization, all; data curation, R.I., M.K. and Y.K., methodology, R.I., M.K. and Y.K.; investigation, R.I., M.K., Y.K., T.O. (Takeru Okada), T.O. (Tatsuro Onomoto), K.T., T.F. and S.K.; writing—original draft preparation, R.I.; writing—review and editing, K.T. and S.K.; visualization, M.K., Y.K., T.O. (Takeru Okada), T.O. (Tatsuro Onomoto) and T.F.; supervision, R.I.; project administration, R.I.; funding acquisition, R.I.

Funding: This work was supported by JSPS KAKENHI Grant Number 15K17482.

Acknowledgments: We wish to acknowledge valuable discussions with Masahiro Okumiya, Toyota Technological Institute, and Nobuyuki Kanayama, Santier Giken Co., Ltd. We are grateful to Masaki Sonoda, Oita Industrial Research Institute, for their technical assistance.

Conflicts of Interest: The authors declare no conflicts of interest.

References

1. Sun, Y.; Bell, T. Plasma surface engineering of low alloy steel. *Mater. Sci. Eng.* **1991**, *140*, 419–434. [CrossRef]
2. Czerwiec, T.; Michel, H.; Bergmann, E. Low-pressure, high-density plasma nitriding: Mechanisms, technology and results. *Surf. Coat. Technol.* **1998**, *108–109*, 182–190. [CrossRef]
3. Rie, K.-T. Recent advances in plasma diffusion processes. *Surf. Coat. Technol.* **1999**, *112*, 56–62. [CrossRef]
4. Li, C.X. Active screen plasma nitriding—An overview. *Surf. Eng.* **2010**, *26*, 135–141. [CrossRef]
5. Nishimoto, A.; Tanaka, T.; Matsukawa, T. Effect of surface deposited layer on active screen plasma nitriding. *Mater. Perform. Charact.* **2016**, *5*, 386–395. [CrossRef]
6. Nishimoto, A.; Matsukawa, T.; Nii, H. Effect of screen open area on active screen plasma nitriding of austenitic stainless steel. *ISIJ Int.* **2014**, *54*, 916–919. [CrossRef]
7. Marcos, G.; Guilet, S.; Cleymand, F.; Thiriet, T.; Czerwiec, T. Stainless steel patterning by combination of micro-patterning and driven strain produced by plasma assisted nitriding. *Surf. Coat. Technol.* **2011**, *205*, S275–S279. [CrossRef]
8. Aizawa, T.; Morita, H.; Wasa, K. Low-temperature plasma nitriding of Mini-/Micro-tools and parts by table-top system. *Appl. Sci.* **2019**, *9*, 1667. [CrossRef]
9. Michel, H.; Czerwiec, T.; Gantois, M.; Ablitzer, D. Progress in the analysis of the mechanisms of ion nitriding. *Surf. Coat. Technol.* **1995**, *72*, 103–111. [CrossRef]
10. Hudis, M. Study of ion-nitriding. *J. Appl. Phys.* **1973**, *44*, 1489. [CrossRef]
11. Matsumoto, O.; Konuma, M.; Kanzaki, Y. Nitriding of titanium in an r.f. discharge II: Effect of the addition of hydrogen to nitrogen on nitriding. *J. Less Common Met.* **1982**, *84*, 157–163. [CrossRef]
12. Sakamoto, Y.; Takaya, M.; Ishii, Y.; Igarashi, S. Surface modified tool fabricated by radical nitriding. *Surf. Coat. Technol.* **2001**, *152–155*. [CrossRef]
13. Lee, I.; Park, I. Microstructure and mechanical properties of surface-hardened layer produced on SKD 61 steel by plasma radical nitriding. *Mater. Sci. Eng.* **2007**, *449*, 890–893. [CrossRef]
14. Tamaki, M.; Tomii, Y.; Yamamoto, N. The role of hydrogen in plasma nitriding: Hydrogen behavior in the titanium nitride layer. *Plasmas Ions* **2000**, *3*, 33–39. [CrossRef]
15. Tahchieva, A.B.; Llorca-Isern, N.; Cabrera, J.-M. Duplex and superduplex stainless steels: microstructure and property evolution by surface modification processes. *Metals* **2019**, *9*, 347. [CrossRef]

16. Almeida, G.F.C.; Couto, A.A.; Reis, D.A.P.; Massi, M.; Da Silva Sobrinho, A.S.; De Lima, N.B. Effect of plasma nitriding on the creep and tensile properties of the Ti-6Al-4V alloy. *Metals* **2018**, *8*, 618. [CrossRef]

17. Ichiki, R.; Nagamatsu, H.; Yasumatsu, Y.; Iwao, T.; Akamine, S.; Kanazawa, S. Nitriding of steel surface by spraying pulsed-arc plasma jet under atmospheric pressure. *Mater. Lett.* **2012**, *71*, 134–136. [CrossRef]

18. Nagamatsu, H.; Ichiki, R.; Yasumatsu, Y.; Inoue, T.; Yoshida, M.; Akamine, S.; Kanazawa, S. Steel nitriding by atmospheric-pressure plasma jet using N_2/H_2 mixture gas. *Surf. Coat. Technol.* **2013**, *225*, 26–33. [CrossRef]

19. Yoshimitsu, Y.; Ichiki, R.; Kasamura, K.; Yoshida, M.; Akamine, S.; Kanazawa, S. Atmospheric-pressure plasma nitriding of titanium alloy. *Jpn. J. Appl. Phys.* **2015**, *54*, 030302. [CrossRef]

20. Maeda, A.; Ichiki, R.; Tomizuka, R.; Nishiguchi, H.; Onomoto, T.; Akamine, S.; Kanazawa, S. Investigation on local formation of expanded austenite phase by atmospheric-pressure plasma jet. In Proceedings of the XXXIII International Conference on Phenomena in Ionized Gases, Estoril, Portugal, 9–14 July 2017; Alves, L.L., Tejero-del-Caz, A., Eds.; p. 169.

21. Sannomiya, R.; Ichiki, R.; Otani, R.; Hanada, K.; Sonoda, M.; Akamine, S.; Kanazawa, S. Investigation on hard-tissue compatibility of TiN surface formed by atmospheric pressure plasma nitriding. *Plasma Fusion Res.* **2018**, *13*, 1306120. [CrossRef]

22. Chiba, S.; Ichiki, R.; Nakatani, T.; Ueno, T.; Kanazawa, S. Development of local evacuation system for inhibiting oxidization in atmospheric-pressure plasma jet nitriding. *Results Phys.* **2019**, *13*, 102131. [CrossRef]

23. Miyamoto, J.; Inoue, T.; Tokuno, K.; Tsutamori, H.; Abraha, P. Surface modification of tool steel by atmospheric-pressure plasma nitriding using dielectric barrier discharge. *Tribol. Online* **2016**, *11*, 460–465. [CrossRef]

24. Kitamura, K.; Ichiki, R.; Tsuru, T.; Akamine, S.; Kanazawa, S. Demonstration of nitriding by dielectric barrier discharge and investigation of treatment range controllability. In Proceedings of the 21st International Conference on Gas Discharges and their Applications, Nagoya, Japan, 11–16 September 2016; Yokomizu, Y., Kojima, H., Eds.; pp. 429–432.

25. Liedtke, D. Gas Nitriding and Nitrocarburizing. In *Wärmebehandlung von Eisenwerkstoffen II: Nitrieren und Nitrocarburieren*; Expert Verlag: Renningen, Germany, 2014.

26. Chiba, M.; Miyamoto, G.; Furuhara, T. Microstructure of pure iron treated by nitriding and quenching process. *J. Jpn. Inst. Metals* **2012**, *76*, 256–264. [CrossRef]

Article

Self-Lubricating PEO–PTFE Composite Coating on Titanium

Limei Ren [1,2,3], Tengchao Wang [1,2], Zhaoxiang Chen [2,3,*], Yunyu Li [2,3] and Lihe Qian [4]

[1] Key Laboratory of Advanced Forging & Stamping Technology and Science (Yanshan University), Ministry of Education of China, Qinhuangdao 066004, China; lmren@ysu.edu.cn (L.R.); tcwang@stumail.ysu.edu.cn (T.W.)

[2] School of Mechanical Engineering, Yanshan University, Qinhuangdao 066004, China; yyl@stumail.ysu.edu.cn

[3] Aviation Key Laboratory of Science and Technology on Generic Technology of Self-lubricating Spherical Plain Bearing, Yanshan University, Qinhuangdao 066004, China

[4] State Key Laboratory of Metastable Materials Science and Technology, Yanshan University, Qinhuangdao 066004, China; lhqian@ysu.edu.cn

* Correspondence: zxchen@ysu.edu.cn (Z.C.); Tel.: +86-335-806-8374

Received: 19 December 2018; Accepted: 30 January 2019; Published: 1 February 2019

Abstract: A self-lubricating plasma electrolytic oxidation–polytetrafluoroethylene (PEO–PTFE) composite coating was successfully fabricated on the surface of commercially pure titanium by a multiple-step method of plasma electrolytic oxidation, dipping and sintering treatment. The microstructure and tribological properties of the PEO–PTFE composite coating were investigated and compared with the PEO TiO_2 coating and the PTFE coating on titanium. Results show that most of the micro-pores of the PEO TiO_2 coating were filled by PTFE and the surface roughness of PEO–PTFE composite coating was lower than that of the PEO TiO_2 coating. Furthermore, the PEO–PTFE composite coating shows excellent tribological properties with low friction coefficient and low wear rate. This study provides an insight for guiding the design of self-lubricating and wear-resistant PEO composite coatings.

Keywords: self-lubricating; composite coating; titanium; plasma electrolytic oxidation (PEO); polytetrafluoroethylene (PTFE)

---

1. Introduction

Titanium and its alloys are lightweight structural metals, exhibiting high strength-to-weight ratio and excellent corrosion resistance. Due to these advantages, they are widely used in many industries, especially the automotive, aerospace and shipping industries [1–3]. However, titanium alloys generally show low surface hardness and poor tribological properties, characterized by severe abrasive wear and adhesive wear, which seriously restricted their applications in field of machinery [4,5]. Thus, developing proper surface modification techniques to improve the tribological properties is a crucial step to expand the application scopes of titanium alloy. At present, the commonly used methods to enhance surface performance of titanium alloys mainly include physical vapor deposition [6], chemical vapor deposition [7], ion implantation [8], thermal spraying [9], plasma electrolytic deposition [10], plasma electrolytic oxidation (PEO) [11], etc. Among these techniques, PEO is a simple and environment-friendly process with rapid deposition of anodic oxide coating on the titanium surface. Moreover, PEO coatings show increased surface hardness and excellent wear resistance. In spite of these advantages, PEO coatings usually take on high surface roughness and porosity with high friction coefficient, which has limited the extensive industrial application of the PEO technique [12]. Therefore, improving the tribological properties of PEO coating, is the key to enlarge its application range. In the previous study, we have explored depositing diamond-like carbon (DLC) on the PEO coating by using the unbalanced magnetron sputtering technique. Although results show that the TiO_2/DLC composite coating exhibits improved tribological properties with low friction coefficient and low wear rate, DLC coating deposition is an expensive and time-consuming process [13]. Furthermore, this technique is not favorable to producing coatings on complex-shaped specimens.

In general, using liquid lubricants can effectively reduce the friction coefficient of the PEO coatings and decrease the wear between the friction pairs. However, in some extreme conditions, such as high temperature and high vacuum, liquid lubricant would volatilize and cause lubrication failure. Solid lubricants commonly used in mechanical lubrication include graphite, hexagonal boron nitride (hBN), molybdenum disulfide (MoS_2), polytetrafluoroethlene (PTFE), etc [4,11,14]. PTFE has good chemical stability and thermal stability, with excellent lubricating property in a relatively wide temperature range and almost all of the ambient atmosphere [15]. In recent years, some researchers have attempted to disperse PTFE particles into the electrolyte and directly incorporated these self-lubricating particles into the anodic coating during the PEO process, but the localized high temperature during the PEO process may cause the decomposition of PTFE. Consequently, it is difficult to control the content and distribution of PTFE in the composite coatings by this technique.

In this paper, with the aim of improving the tribological properties of the PEO coating, especially the lubricating property, the PEO coating was used as the substrate and then the solid lubricant PTFE was deposited directly on the surface of the PEO coating by dipping and sintering treatment to fabricate a PEO–PTFE composite coating. The surface of the PEO coating is characterized with lots of micro-pores and micro-cracks which ensured the probability to deposit the small sized solid lubricant into these micro-pores on the PEO coating. The microstructure and tribological properties of the PEO–PTFE composite coating were investigated using scanning electron microscopy (SEM), energy-dispersive X-ray spectroscopy (EDS), Raman, surface profiler and ball-on-plate tribometer test and then compared with the PEO TiO_2 coating and the Ti-PTFE coating.

2. Materials and Methods

2.1. Preparation of Coatings

The commercially pure titanium was used as the substrate for preparing the PEO coating, and the titanium plate was cut into samples with a size of $25 \times 70 \times 2$ mm. The chemical composition of the titanium substrate were Fe ≤ 0.30 %, Si ≤ 0.15 %, O ≤ 0.20 %, C ≤ 0.10 %, N ≤ 0.05 %, H ≤ 0.015 % and Ti balance. In preparation for PEO process, all samples were ground with SiC abrasive papers (from 150-grit to 1500-grit) and cleaned ultrasonically with 95% alcohol for 30 min. Then, all samples were cleaned with deionized water and air dried. A pulsed asymmetric bipolar AC power supply (MAO120HD-III, Xi'an University of Technology, China) was employed for the PEO process. The PEO of titanium was carried out under the mode of constant voltage and the parameters setting of PEO is listed in Table 1. The electrolyte used in the PEO process consisted of Na_2SiO_3 (15 g/L), Na_3PO_4 (10 g/L) and NaOH (1 g/L). During the PEO treatment, titanium samples and stainless steel plates were used as the anode and cathode, respectively. A circulating cooling system was used to keep the temperature of the electrolyte below 40 °C. After the PEO treatment, the coated samples were cleaned with deionized water and dried at ambient temperature.

Table 1. The parameters setting of plasma electrolytic oxidation power supply.

Parameter	Voltage (V)	Frequency (Hz)	Duty Ratio (%)	Processing Time (min)
Forward	420	500	35	25
Reverse	50	500	15	

After the PEO treatment, the samples were subjected to dipping and sintering treatment to fabricate the PEO–PTFE composite coating. The PEO samples were immersed in PTFE dispersion

(Shanghai Aladdin Bio-Chem Tech Co. LTD, Shanghai, China) heated to 50 °C and held for 20 min in a thermostat water bath, and then put into a muffle furnace (TESE, RXF1400-5-12, Shanghai, China). Figure 1 shows the process of sintering treatment for PEO–PTFE composite coating. Firstly, all coated samples placed in the muffle furnace were heated from room temperature to 150 °C with a heating rate of 5 °C/min, keeping the temperature at 150 °C for 20 min. After that, the temperature was increased to 350 °C with a heating rate of 5 °C/min, holding at the temperature of 350 °C for 20 min. Finally, all samples were cooled to room temperature inside the furnace. Following the same procedure as mentioned above, the PTFE coating directly deposited on the titanium was fabricated.

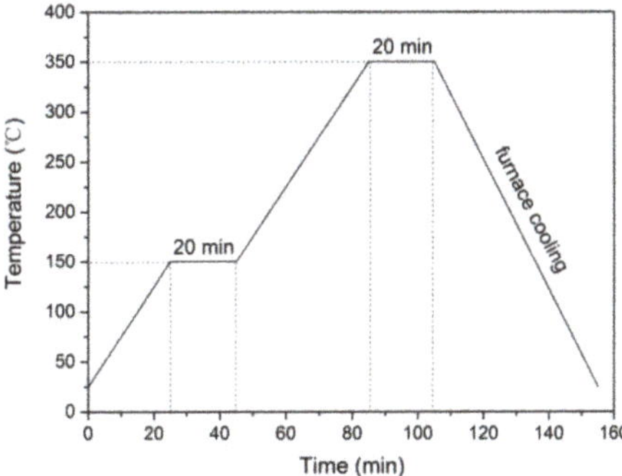

Figure 1. Sintering treatment process.

2.2. Characterization of Coatings

The surface morphologies of the coatings were studied by using scanning electron microscopy (SEM, SIGMA-500, ZEISS, Jena, Germany), equipped with EDS. The EDS was used to analyze the elementary composition of the PEO–PTFE composite coating. Raman spectra of the coatings were obtained with an Nd-YAG solid state laser (wavelength 532.0 nm) through the 50x objective lens of a Raman microspectrometer (Horiba, Kyoto, Japan). The surface roughness of the coatings was examined by a confocal microscope (Conscan Profilometer, Anton Paar Tritec SA, Peseux, Switzerland). The tribological properties of coatings were measured at room temperature under dry sliding condition by using a ball-on-plate reciprocating tribometer (Tribometer, Anton-Paar Tritec SA, Peseux, Switzerland). And the GCr15 stainless steel ball, with a diameter of 3 mm, was used as grinding material in the test. The friction coefficient and wear rate of the coatings was measured at different loads (2~8 N) with a sliding distance of 100 m. After the friction and wear tests, the wear tracks were measured by a surface profiler (MarSurf, Mahr, Göttingen, Germany). The wear rate of samples was calculated according to the formula: $K = (S \cdot l)/(F \cdot L)$ [16], where S is the cross-sectional area of wear track (mm^2), l is the length of wear track (mm), F is the applied load (N), L is the sliding distance (m).

3. Results and Discussion

3.1. Surface Morphology and Compositions

Figure 2a–d illustrates the morphologies of titanium substrate, Ti-PTFE coating, PEO TiO$_2$ coating and PEO–PTFE composite coating. The SEM image of titanium substrate ground by sandpapers is shown in Figure 2a, from which lots of plough marks can be observed. This phenomenon reflects poor wear resistance of the commercially pure titanium. Figure 2b displays the morphology of the PTFE coating deposited directly on the surface of titanium substrate. It can be seen that the PTFE coating did not cover the substrate very well and peeled off at some areas, indicating the poor bonding between the PTFE coating and the relatively smooth titanium substrate. The typical surface morphology of PEO TiO$_2$ coating fabricated in the silicate-phosphate electrolytic solution is presented in Figure 2c.

It can be seen that the PEO TiO$_2$ coating has a porous and rough surface structure, characterized with a large number of crater-like micro-pores and micro-protrusions. These micro-pores were formed by the plasma micro-arc discharges, which resulted in local high temperature and high pressure, causing the molten materials to erupt from the micro-arc discharge channels. Then, the erupted molten oxides solidified and accumulated around the micro-pores, leading to the formation of micro-protrusions. Although these surface structures led to a high surface roughness, this provided the possibility of depositing small sized solid lubricant into micro-pores and around micro-protrusions. In addition, the cross-sectional morphology of PEO TiO$_2$ coating is also shown in Figure 2c. It can be seen that the oxide coating adhered tightly to the substrate and the average thickness of the oxide coating is around 10 μm.

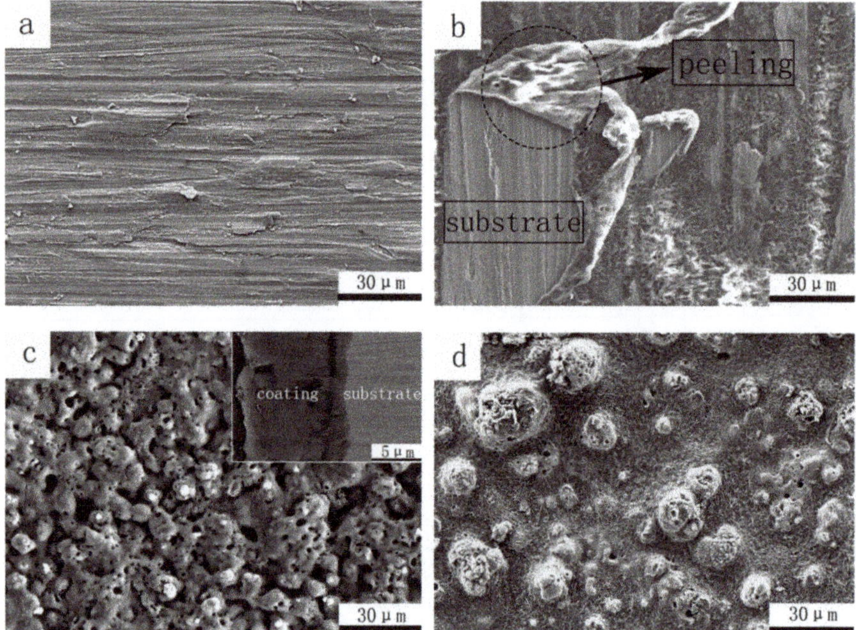

Figure 2. Morphologies of (**a**) titanium substrate; (**b**) Ti-polytetrafluoroethylene (PTFE) coating; (**c**) plasma electrolytic oxidation (PEO) TiO$_2$ coating; (**d**) PEO–PTFE composite coating.

Figure 2d shows the surface morphology of PEO–PTFE composite coating. Compared with Figure 2c, it can be observed that a great deal of micro-pores on the PEO TiO$_2$ coating had been filled with PTFE and the porosity of the PEO TiO$_2$ coating decreased significantly. During the process of sintering treatment, the sample was heated to 350 °C, which exceeded the melting point of PTFE materials [17]. As the PTFE has a dynamic viscosity in the melting state, the melted PTFE filled the pores of PEO and formed a continuous composite PEO–PTFE coating [18–20]. Obviously, the rough and porous PEO surface enhanced the crosslinking and bonding performance of PTFE materials to the bottom of the TiO$_2$ coating.

The EDS elemental maps of the PEO–PTFE composite coating is shown in Figure 3, which are obtained from the EDS scanning of Figure 2d. F and C elements were detected in the PEO–PTFE composite coating, indicating that PTFE materials deposited successfully on the PEO coating. From the distribution of F and C elements, it can be found that the most area of the PEO coating was covered by PTFE materials, except for the locations where the PEO oxides highly accumulated. This phenomenon is because of the fluidity of PTFE materials during the dipping and sintering processing. The mechanical interlinking formed between the porous PEO coating and the continuous PTFE materials after the sintering treatment. Such mechanical interlinking ensured a good bonding performance of the PEO–PTFE composite coating. In contrast to the single PEO TiO_2 coating, this PEO–PTFE composite coating maintained the good wear-resistance of the ceramic PEO component while its PTFE component played the role as the friction-reducing lubricant. As reported in a recent study, a self-lubricating PEO coating was fabricated on AZ91 magnesium alloy via the in-situ incorporation of PTFE particles [21]. Results showed that the in-situ PTFE incorporation into the growing PEO coating resulted in non-uniform PTFE distribution and some PTFE-enriched ridge-like protrusions were formed on the coating surface. When the incorporation time was sufficient, the PTFE-enriched protrusions formed can act as lubricant reservoirs, leading to a low and stable friction coefficient. The present PEO–PTFE composite coating aimed to achieve uniform PTFE distribution by using the dipping and sintering treatments, ensuring even and sufficient PTFE lubricant supply during the friction and wear test.

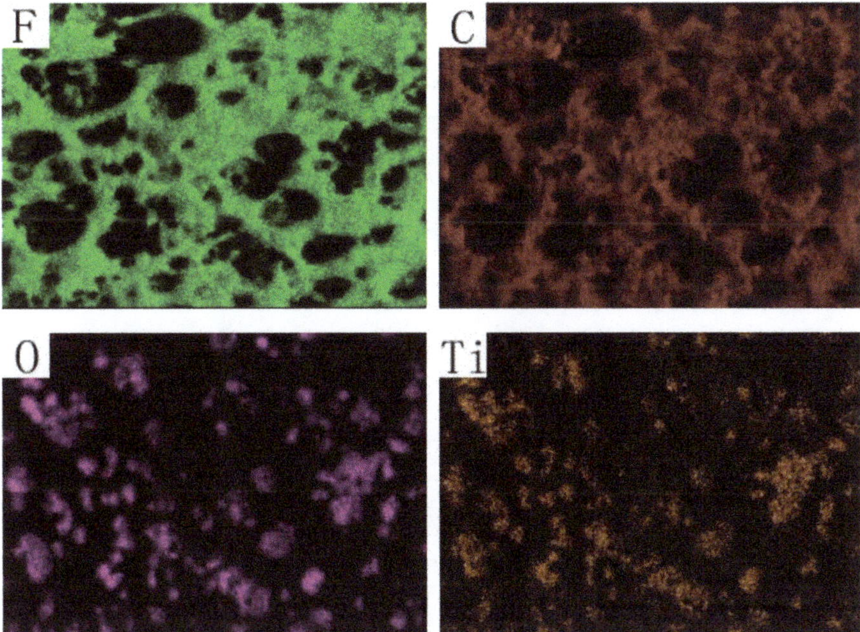

Figure 3. Energy-dispersive X-ray spectroscopy (EDS) elemental maps of PEO–PTFE composite coating.

The Raman spectra of PEO TiO_2 coating and PEO–PTFE composite coating are shown in Figure 4. Recorded Raman spectra were made in the frequency range from 100 to 1500 cm^{-1}. It can be seen from the Raman spectrum of PEO TiO_2 coating that bands shift located at 129 (E_g) and 498 (A_{1g}) cm^{-1} positions originate from different modes of anatase TiO_2 [22]. Bands shift located at 423 (E_g) and 591 (A_{1g}) cm^{-1} positions originate from different modes of rutile TiO_2. It can be inferred from Raman spectra of the PEO coating that there are two TiO_2 crystal structures with anatase and rutile on the surface of the coating. According to the Raman spectrum of PEO–PTFE composite coating, one can see that some characteristic peaks of PTFE can be detected. The Raman peaks at 271, 364, 711 and 1360 cm^{-1} are related to the different vibrational modes of CF_2 groups, whereas bands shift located at 1218 and 1282 cm^{-1} positions come from the bands C-C and CF, respectively [23]. The Raman result is consistent with the previous EDS analysis result, indicating the successful deposition of PTFE on the PEO coating.

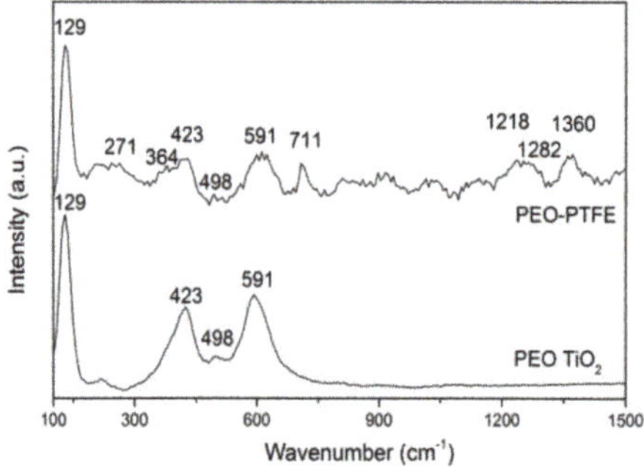

Figure 4. Raman spectra of the PEO TiO$_2$ coating and PEO–PTFE composite coating.

Figure 5 presents the surface roughness (Ra) of the titanium substrate, Ti-PTFE coating, PEO TiO$_2$ coating, and PEO–PTFE composite coating. Compared with the titanium substrate, the roughness of Ti-PTFE coating slightly increased. This is because the Ti-PTFE coating deposited directly on the smooth titanium substrate was easy to peel off (see Figure 2b), causing a rougher surface than the titanium substrate. The roughness of the PEO TiO$_2$ coating turned out to be the highest due to the existence of micro-pores and micro-protrusions on its surface (see Figure 2c). PTFE polymer materials deposited into the porous TiO$_2$ coating (see Figure 2d) and formed the composite coating. Therefore, the roughness of the PEO–PTFE composite coating decreased significantly compared with the PEO TiO$_2$ coating.

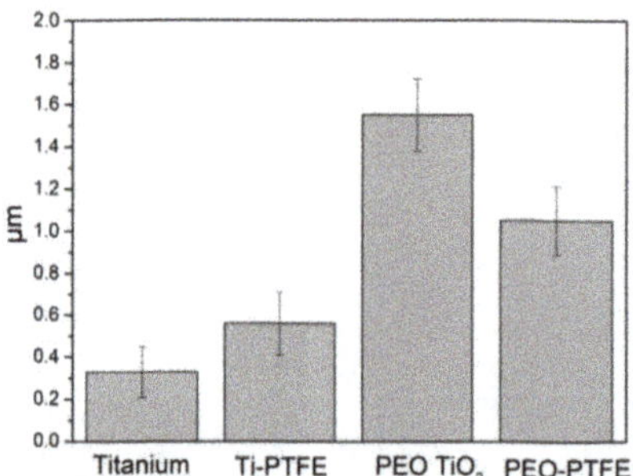

Figure 5. Surface roughness (Ra) of titanium substrate, Ti-PTFE coating, PEO TiO$_2$ coating and PEO–PTFE composite coating.

3.2. Tribological Behaviors

The tribological tests were performed at room temperature using a ball-on-plate reciprocating tribometer. Figure 6a shows the relationship between the friction coefficient and sliding distance of the titanium substrate, Ti-PTFE coating, PEO TiO$_2$ coating and PEO–PTFE composite coating. From the friction coefficient curve of titanium substrate, it can be seen that the value remained about 0.5 after a slight fluctuation at the initial stage of the test. In the case of PEO TiO$_2$ coating, this curve exhibited significant fluctuation before the sliding distance of about 20 m. After that, the friction coefficient gradually decreased to 0.65 and remained stable until the end of the test. This phenomenon was in connection with the unique structure of PEO TiO$_2$ coating. The PEO coating typically has a rough and porous outer layer and a dense inner layer [13]. The porous outer layer featured with high surface roughness and varied internal microstructure, which contributed to the high and unstable friction coefficient at the initial stage of the test. With the gradual wearing-off of the porous outer layer, the GCr15 ball began to contact and slide against the dense inner layer of PEO TiO$_2$ coating, resulting in the decreased and relatively stable friction coefficient.

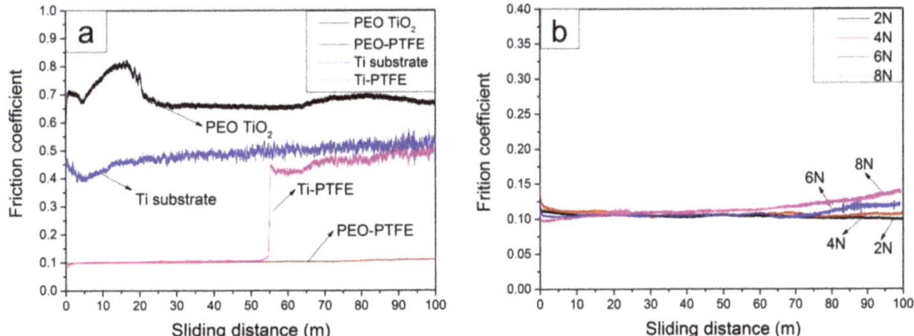

Figure 6. Friction coefficient of different samples against GCr15 after a sliding distance of 100 m (**a**) titanium substrate, Ti-PTFE coating, PEO TiO$_2$ coating and PEO–PTFE composite coating under the load of 4 N; (**b**) PEO–PTFE composite coating under different loads.

Comparing the friction coefficient curves of the Ti-PTFE coating and the PEO–PTFE composite coating in Figure 6a, it can be seen that the friction coefficient of PEO–PTFE composite coating remained stable at 0.1 during the whole process of the test. For the Ti-PTFE coating, the friction coefficient was about 0.1 from the beginning until the sliding distance of 55 m and then increased sharply to about 0.4, which was close to the coefficient of titanium substrate. This result indicated that the Ti-PTFE coating had been worn off after a certain sliding distance. Afterwards, the friction behavior occurred between the grinding ball GCr15 and titanium substrate. As shown in Figure 2b, the PTFE coating exhibited poor adhesion to the titanium substrate. Therefore, it was easily damaged and peeled off in the process of reciprocating friction and wear. In the case of the PEO–PTFE composite coating, its long-term low friction coefficient proved that depositing of PTFE materials can effectively improve the friction property of PEO TiO$_2$ coating. Also, it is demonstrated that the self-lubricating friction behavior occurred between the grinding pairs.

Figure 6b shows the relationship between the friction coefficient and sliding distance of the PEO–PTFE coating under different loads. It can be observed that when the load was 2 N and 4 N, the friction coefficient was stable at about 0.1. Under the higher load of 6 N and 8 N, the friction coefficient of the self-lubricating PEO–PTFE composite coating increased slightly with the increase of the sliding distance, but was still lower than 0.15. The friction coefficient curves of the self-lubricating PEO–PTFE composite coating under different loads consistently appeared to be low and stable, indicating that the self-lubricating PEO–PTFE composite coating has a good load-bearing capacity. Recently, Wang et al. constructed a lubricant composite coating on Ti6Al4V alloy using micro-arc oxidation and grafting hydrophilic polymer [24]. Results showed that the composite coating exhibited the low friction coefficient and favorable wear resistance in water under a low contact stress of 1.52 MPa. It was explained that the hydrophilic polymer formed a hydrated lubricating layer through the interaction with water and the TiO$_2$ ceramic layer provided the resistance to wear.

Figure 7a–c presents the SEM images of wear tracks of the Ti-PTFE coating, PEO TiO$_2$ coating and the PEO–PTFE composite coating after a sliding distance of 100 m under the normal load of 4 N. The cross-sectional profiles of wear tracks are shown in Figure 7d. From the wear track profiles and worn morphologies of the coated samples, it can be noticed that the wear track of Ti-PTFE coating was narrower and deeper than the PEO TiO$_2$ coating. Due to the loose adhesion of PTFE coating to titanium substrate, the PTFE coating was easily peeled off and damaged. After the PTFE coating was worn out, the titanium substrate was exposed to the grinding ball GCr15. Therefore, the poor wear resistance of titanium substrate resulted in a deeper wear track. For the PEO TiO$_2$ coating, it can be seen that although the width of the wear track was large, the overall wear track was shallow. The ceramic PEO TiO$_2$ coating has the advantage of high hardness and excellent wear resistance. Consequently, a lot of wear and tear of the GCr15 counterpart occurred with the proceeding wear test. Therefore, the contact areas between the grinding pairs increased, leading to the increase of the wear track width. For the PEO–PTFE composite coating, it can be seen from Figure 7c that the wear track was obviously shallower and narrower than both the Ti-PTFE coating and PEO TiO$_2$ coating. This is because the deposition of PTFE polymer materials on the PEO TiO$_2$ coating significantly improved the tribological properties of the PEO TiO$_2$ coating. Firstly, the deposition PTFE materials in the PEO micro-pores or around the PEO micro-protrusions resulted in decreased the surface roughness and thus increased contact area and decreased contact stress between the friction pair, which avoided the rapid fracture and wearing-off of PEO outer layer to some extent. Secondly, the existence of self-lubricating PTFE component in the composite coating decreased the friction force and kept the GCr15 counterpart from severe wear.

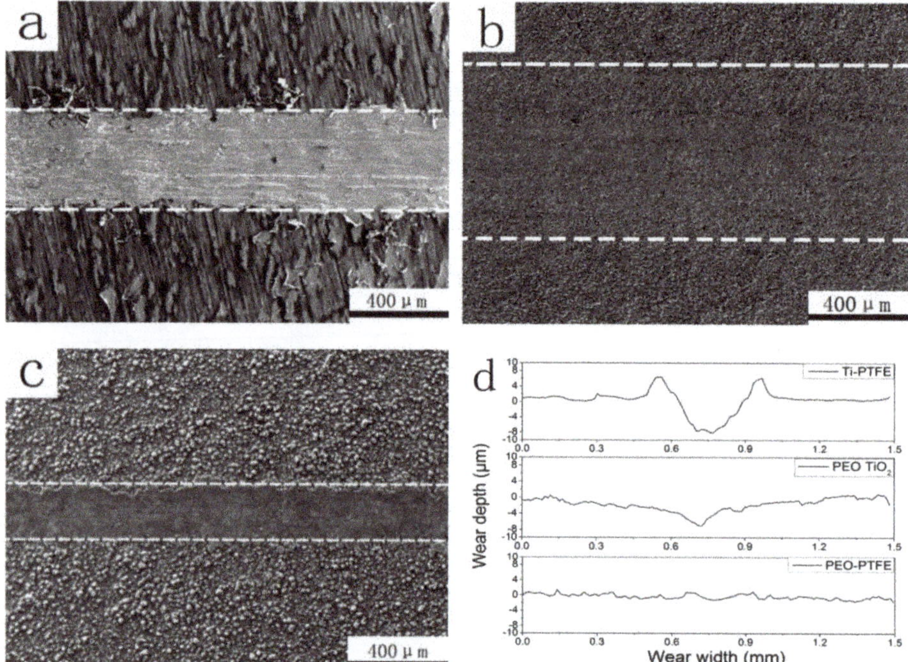

Figure 7. Scanning electron microscope (SEM) images and cross-sectional profiles of wear tracks after a sliding distance of 100 m under the 4 N load of the (**a**) Ti-PTFE coating; (**b**) PEO TiO$_2$ coating; (**c**) PEO–PTFE composite coating; and (**d**) cross-sectional profiles of wear tracks.

Figure 8 shows the micro worn morphologies of the PEO TiO$_2$ coating and the self-lubricating PEO–PTFE composite coating after a sliding distance of 100 m under the 4 N loading. The element composition of wear tracks were examined by EDS and the results are listed in Table 2. For the worn track of PEO TiO$_2$ coating (Figure 8a), it can be seen that there were still a great deal micro-pores after the tribo-test, indicating high porosity both in its surface and interior. The existence of micro-pores and debris greatly weakened the strength and toughness of the PEO TiO$_2$ coating. It can be observed that there were two kinds of typical worn surface morphologies of the PEO TiO$_2$ coating, as marked by point 1 and point 2 in Figure 8a. The first worn morphology (see point 1 in Figure 8a) was relatively smooth. A lot of Fe and O elements as well as a small quality of Cr element were detected (see Table 2), indicating that materials were transferred from the GCr15 ball to the PEO coating surface during the wear process. The second worn morphology (see point 2 in Figure 8a) was relatively rough. The brittle and porous surface layer of the PEO coating was crushed under the normal loading and formed lots of oxide debris at the initial stage of the tribo-test, which caused abrasive wear between the PEO coating and GCr15 ball during the subsequent tribo-test. Therefore, lots of O, Ti and Si elements and small amounts of Fe elements derived from GCr15 ball were detected (see Table 2).

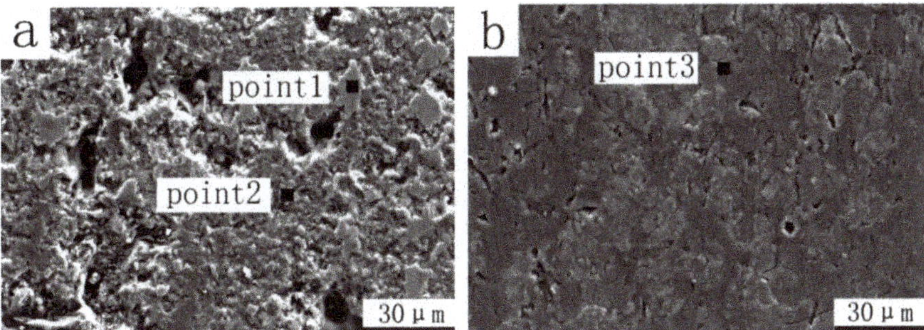

Figure 8. Micro worn morphologies after a sliding distance of 100 m under the load of 4 N: (**a**) PEO TiO$_2$ coating; (**b**) the self-lubricating PEO–PTFE composite coating.

Table 2. Element composition of the PEO TiO$_2$ coating and PEO–PTFE composite coating worn track after a sliding distance of 100 m under 4 N in Figure 8.

Element (wt %)	O	C	Fe	F	Si	Ti	Cr	P	Na
Point 1	33.3	5.7	55.8	-	1.2	3.3	0.7	-	-
Point 2	47	7.7	10.1	-	17.6	16.2	-	1.4	-
Point 3	34.2	20.5	0.3	20.2	20.1	1.9	-	1.6	1.2

As shown in Figure 8b, the worn surface morphology of the self-lubricating PEO–PTFE composite coating was obviously different from the PEO TiO$_2$ coating. There were only a small number of micro-pores being observed on the worn surface, indicating a compact worn surface of the self-lubricating PEO–PTFE composite formed by repeated crushing of PTFE polymer materials and PEO TiO$_2$ coating. In addition, a lot of F and C elements (see point 3 in Figure 8b and Table 2) were detected on the worn surface, demonstrating the formation of self-lubricating film. As PTFE polymer materials deposited into the PEO TiO$_2$ coating, the porous surface was covered by the self-lubricating materials. The micro-pores were sealed by the solid lubricant. In the process of friction, a lubricating film would be formed on the friction contact surface. As a result, the sliding pairs in the contact surface were changed from the steel ball versus the composite coating to PTFE transfer film on the surface of the steel ball versus PTFE materials. Because of the poor adhesion between the PTFE transfer film and the steel ball, the transfer film usually fell off under the obstacle of the abrasive particle [25]. But the self-lubricating materials stored in the micro-pores provided continuous supply for the formation of the self-lubricating film. As the sliding distance increased, the PEO TiO$_2$ component played a role as a wear-resistant reinforcing phase and supported the self-lubricating film.

Figure 9 shows the wear rate of all samples against the counterpart GCr15 ball after the sliding test under the normal load of 4 N and a sliding distance of 100 m. During the process of sliding, the titanium substrate sample suffered from severe wear and tear, mainly due to its low hardness. Therefore, the wear rate value of the titanium substrate is large, which is 99.27 $\times$ 10^{-5} mm$^3\cdot$N$^{-1}\cdot$m^{-1}. For the Ti-PTFE sample, the average wear rate was 28.39 $\times$ 10^{-5} mm$^3\cdot$N$^{-1}\cdot$m^{-1}, smaller than that of the titanium substrate sample, because there was self-lubricating friction behavior at the beginning stage of the tribo-test (see Figure 6a). When the PTFE coating was worn off and the self-lubricating film failed, the bottom titanium substrate began to be subject to wear. From Figure 9, it can be seen that the wear rate of the PEO TiO$_2$ coating (7.02 $\times$ 10^{-5} mm$^3\cdot$N$^{-1}\cdot$m^{-1}) was smaller than that of both the titanium substrate and Ti-PTFE samples. The PEO TiO$_2$ coating sample has higher hardness than Ti and Ti-PTFE samples, leading to the decrease of wear rate. The wear rate of the self-lubricating PEO–PTFE composite coating was 1.34 $\times$ 10^{-5} mm$^3\cdot$N$^{-1}\cdot$m^{-1}, which is the smallest among all the samples. As the PTFE was integrating into the PEO coating and filling into the micro-pores, the high hardness of PEO coating and the good lubrication of PTFE combined to increase the wear resistance and decrease the friction coefficient.

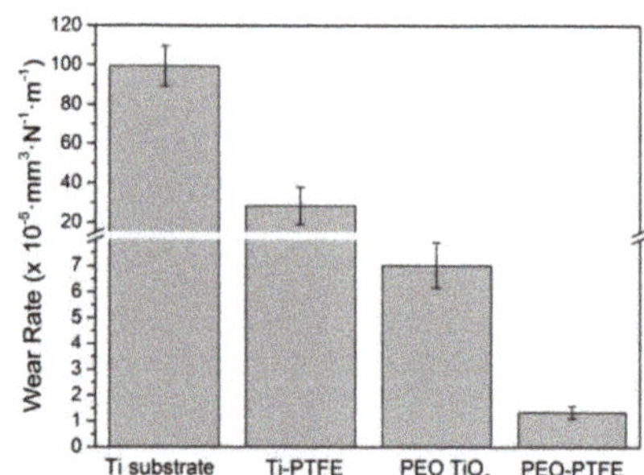

Figure 9. Wear rate of the titanium substrate, Ti-PTFE, PEO TiO$_2$ coating and the self-lubricating PEO–PTFE composite coating under the load of 4 N and the sliding distance of 100 m.

4. Conclusions

The self-lubricating PEO–PTFE composite coating was successfully fabricated on the titanium surface by the combined methods of plasma electrolytic oxidation, dipping and sintering. Most micro-pores of the PEO TiO$_2$ coating were effectively filled with PTFE and the surface roughness of the PEO TiO$_2$ coating decreased significantly. The fabricated PEO–PTFE composite coating, integrating the advantages of wear resistance of the PEO TiO$_2$ coating and the self-lubrication of PTFE polymer materials, exhibited a low friction coefficient and wear rate. The friction coefficient of the PEO–PTFE composite coating was much smaller than that of PEO TiO$_2$ coating and remained stable (around 0.1) under different loads. The PEO–PTFE composite coating was more durable than the single PTFE coating and its wear rate was about 5 times lower than that of the PEO TiO$_2$ coating.

Author Contributions: Z.C. and L.R. conceived and designed the experiments; T.W. and Y.L. performed the experiments; L.R. and T.W. analyzed the data and wrote the paper; Z.C. supervised the work; L.Q. contributed analysis tools.

Funding: This research was supported by Natural Science Foundation (E2016203270) and Returned Overseas Chinese Talents Foundation (CL201726) of Hebei Province, and the Fundamental Research Foundation (020000904) and Doctoral Foundation (B942) of Yanshan University.

Conflicts of Interest: The authors declare no conflict of interest.

References

1. Diamanti, M.V.; Sebastiani, M.; Mangione, V.; Del Curto, B.; Pedeferri, M.P.; Bemporad, E.; Cigada, A.; Carassiti, F. Multi-step anodizing on Ti6Al4V components to improve tribomechanical performances. *Surf. Coat. Technol.* **2013**, *227*, 19–27. [CrossRef]
2. Meng, F.; Li, Z.; Liu, X. Synthesis of tantalum thin films on titanium by plasma immersion ion implantation and deposition. *Surf. Coat. Technol.* **2013**, *229*, 205–209. [CrossRef]
3. Yuan, X.; Tan, F.; Xu, H.; Zhang, S.; Qu, F.; Liu, J. Effects of different electrolytes for micro-arc oxidation on the bond strength between titanium and porcelain. *J. Prosthodont. Res.* **2017**, *61*, 297–304. [CrossRef] [PubMed]
4. Ao, N.; Liu, D.; Wang, S.; Zhao, Q.; Zhang, X.; Zhang, M. Microstructure and tribological behavior of a TiO$_2$/hBN composite ceramic coating formed via micro-arc oxidation of Ti–6Al–4V alloy. *J. Mater. Sci. Technol.* **2016**, *32*, 1071–1076. [CrossRef]
5. Aliofkhazraei, M.; Sabour Rouhaghdam, A.; Shahrabi, T. Abrasive wear behaviour of Si$_3$N$_4$/TiO$_2$ nanocomposite coatings fabricated by plasma electrolytic oxidation. *Surf. Coat. Technol.* **2010**, *205*, S41–S46. [CrossRef]

6. Marin, E.; Offoiach, R.; Regis, M.; Fusi, S.; Lanzutti, A.; Fedrizzi, L. Diffusive thermal treatments combined with PVD coatings for tribological protection of titanium alloys. *Mater. Des.* **2016**, *89*, 314–322. [CrossRef]

7. Zhu, Y.; Wang, W.; Jia, X.; Akasaka, T.; Liao, S.; Watari, F. Deposition of TiC film on titanium for abrasion resistant implant material by ion-enhanced triode plasma CVD. *Appl. Surf. Sci.* **2012**, *262*, 156–158. [CrossRef]

8. Jin, G.; Cao, H.; Qiao, Y.; Meng, F.; Zhu, H.; Liu, X. Osteogenic activity and antibacterial effect of zinc ion implanted titanium. *Colloids Surf. B Biointerfaces* **2014**, *117*, 158–165. [CrossRef]

9. Daram, P.; Banjongprasert, C.; Thongsuwan, W.; Jiansirisomboon, S. Microstructure and photocatalytic activities of thermal sprayed titanium dioxide/carbon nanotubes composite coatings. *Surf. Coat. Technol.* **2016**, *306*, 290–294. [CrossRef]

10. Yang, X.; Jiang, Z.P.; Ding, X.F.; Hao, G.J.; Liang, Y.F.; Lin, J.P. Influence of solvent and electrical voltage on cathode plasma electrolytic deposition of Al_2O_3 antioxidation coatings on Ti-45Al-8.5Nb alloy. *Metals* **2018**, *8*, 308. [CrossRef]

11. Lu, X.; Mohedano, M.; Blawert, C.; Matykina, E.; Arrabal, R.; Kainer, K.U.; Zheludkevich, M.L. Plasma electrolytic oxidation coatings with particle additions—A review. *Surf. Coat. Technol.* **2016**, *307*, 1165–1182. [CrossRef]

12. Wang, S.; Zhao, Q.; Liu, D.; Du, N. Microstructure and elevated temperature tribological behavior of TiO_2/Al_2O_3 composite ceramic coating formed by microarc oxidation of Ti6Al4V alloy. *Surf. Coat. Technol.* **2015**, *272*, 343–349. [CrossRef]

13. Chen, Z.X.; Ren, X.P.; Ren, L.M.; Wang, T.C.; Qi, X.W.; Yang, Y.L. Improving the tribological properties of spark-anodized titanium by magnetron sputtered diamond-like carbon. *Coatings* **2018**, *8*, 83. [CrossRef]

14. Zhang, Y.; Li, C.; Jia, D.; Zhang, D.; Zhang, X. Experimental evaluation of the lubrication performance of MoS_2/CNT nanofluid for minimal quantity lubrication in Ni-based alloy grinding. *Int. J. Mach. Tools Manuf.* **2015**, *99*, 19–33. [CrossRef]

15. Zhang, R.; Zhao, J.; Liang, J. A novel multifunctional PTFE/PEO composite coating prepared by one-step method. *Surf. Coat. Technol.* **2016**, *299*, 90–95. [CrossRef]

16. Kavimani, V.; Soorya Prakash, K.; Thankachan, T. Surface characterization and specific wear rate prediction of r-GO/AZ31 composite under dry sliding wear condition. *Surf. Interfaces* **2017**, *6*, 143–153. [CrossRef]

17. Khatipov, S.A.; Kabanov, S.P.; Konova, E.M.; Ivanov, S.A.; Serov, S.A. Effect of PTFE irradiation above the melting point on ITS porosity. *Radiat. Phys. Chem.* **2012**, *81*, 273–277. [CrossRef]

18. Gamboni, O.C.; Riul, C.; Billardon, R.; Bose Filho, W.W.; Schmitt, N.; Canto, R.B. On the formation of defects induced by air trapping during cold pressing of PTFE powder. *Polymer* **2016**, *82*, 75–86. [CrossRef]

19. Bai, D.Y.; Bai, H.W.; Fu, Q. Recent progress on sintering molding of polymers. *Polym. Bull.* **2017**, *10*, 13–22. [CrossRef]

20. Huang, Q.L.; Xiao, C.F.; Hu, X.Y.; Bian, L.N. Preparation and properties of poly (Tetrafluoroethylene) membrane. *Polym. Mater. Sci. Eng.* **2010**, *26*, 123–126. [CrossRef]

21. Chen, Y.; Lu, X.; Blawert, C.; Zheludkevich, M.L.; Zhang, T.; Wang, F. Formation of self-lubricating PEO coating via in-situ incorporation of PTFE particles. *Surf. Coat. Technol.* **2018**, *337*, 379–388. [CrossRef]

22. Chen, Z.X.; Zhou, K. Surface morphology, phase structure and property evolution of anodized titanium during water vapor exposure. *Surf. Coat. Technol.* **2015**, *263*, 61–65. [CrossRef]

23. Wyszkowska, E.; Leśniak, M.; Kurpaska, L.; Prokopowicz, R.; Jozwik, I.; Sitarz, M.; Jagielski, J. Functional properties of poly(tetrafluoroethylene) (PTFE) gasket working in nuclear reactor conditions. *J. Mol. Struct.* **2018**, 306–311. [CrossRef]

24. Wang, K.; Xiong, D. Construction of lubricant composite coating on Ti6Al4V alloy using micro-arc oxidation and grafting hydrophilic polymer. *Mater. Sci. Eng. C* **2018**, *90*, 219–226. [CrossRef]

25. Gong, D.L.; Xue, Q.J. Transfer and adhesive wear of polytetrafluroethylene and its composite materials. *J. Solid Lubr.* **1990**, *10*, 73–83. [CrossRef]

 *metals*

Article

Time Evolution Characterization of Atmospheric-Pressure Plasma Jet (APPJ)-Synthesized Pt-SnO$_x$ Catalysts

Chia-Chun Lee [1], Tzu-Ming Huang [1,2], I-Chun Cheng [3,4,*], Cheng-Che Hsu [5,*] and Jian-Zhang Chen [1,2,*]

[1] Graduate Institute of Applied Mechanics, National Taiwan University, Taipei 10617, Taiwan; r05543015@ntu.edu.tw (C.-C.L.); r06543009@ntu.edu.tw (T.-M.H.)

[2] Advanced Research Center for Green Materials Science and Technology, National Taiwan University, Taipei 10617, Taiwan

[3] Graduate Institute of Photonics and Optoelectronics, National Taiwan University, Taipei 10617, Taiwan

[4] Department of Electrical Engineering, National Taiwan University, Taipei 10617, Taiwan

[5] Department of Chemical Engineering, National Taiwan University, Taipei 10617, Taiwan

[*] Correspondence: iccheng@ntu.edu.tw (I.-C.C.); chsu@ntu.edu.tw (C.-C.H.); jchen@ntu.edu.tw (J.-Z.C.); Tel.: +886-2-33669648 (I.-C.C.); +886-2-33663034 (C.-C.H.); +886-2-33665694 (J.-Z.C.)

Received: 17 August 2018; Accepted: 29 August 2018; Published: 1 September 2018

Abstract: We characterize the time evolution ($\leq$120 s) of atmospheric-pressure plasma jet (APPJ)-synthesized Pt-SnO$_x$ catalysts. A mixture precursor solution consisting of chloroplatinic acid and tin(II) chloride is spin-coated on fluorine-doped tin oxide (FTO) glass substrates, following which APPJ is used for converting the spin-coated precursors. X-ray photoelectron spectroscopy (XPS) indicates the conversion of a large portion of metallic Pt and a small portion of metallic Sn (most Sn is in oxidation states) from the precursors with 120 s APPJ processing. The dye-sensitized solar cell (DSSC) efficiency with APPJ-synthesized Pt-SnO$_x$ CEs is improved greatly with only 5 s of APPJ processing. Electrochemical impedance spectroscopy (EIS) and Tafel experiments confirm the catalytic activities of Pt-SnO$_x$ catalysts. The DSSC performance can be improved with a short APPJ processing time, suggesting that a DC-pulse nitrogen APPJ can be an efficient tool for rapidly synthesizing catalytic Pt-SnO$_x$ counter electrodes (CEs) for DSSCs.

Keywords: atmospheric pressure plasma jet; platinum; tin oxide; dye-sensitized solar cells; chloroplatinic acid; tin chloride

---

1. Introduction

Atmospheric-pressure plasma (APP) technology is operated without using a vacuum chamber and associated pumping system. It is therefore considered a cost-effective manufacturing tool. Recent developments have resolved stability and arcing problems, making APP technology promising for industrial applications. Traditional APP sources include transferred arc, corona discharge, dielectric barrier discharge, and atmospheric pressure plasma jet (APPJs) [1,2]. APPs with various heavy particle temperatures and charge densities can be produced by using different excitation methods and electrode configuration designs. The synergy between the reactive plasma species and heat can promote rapid chemical reactions during material processing [3–6]. APPs have been used for processing various types of materials, such as carbon nanotubes [3,7,8] and reduced graphene oxides [9–11]. Applications of APPs for surface cleaning or modification [12–14], deposition of metal oxides [15,16], and syntheses of metal compounds from liquid precursors [6,17,18] have been extensively investigated. Metals and metal oxides are common catalysts [19–24]. APPs also have been used for syntheses and post-treatments of catalysts [25].

In 1991, Grätzel et al. reported a great breakthrough of DSSCs [26], and since then, dye-sensitized solar cells (DSSCs) have been extensively investigated. A conventional DSSC consists of a dye-adsorbed photoanode, an electrolyte, and a counter electrode (CE). A catalytic CE is used for reducing triiodide into iodine in the electrolyte. Generally, Pt is the most commonly used CE material in DSSC, owing to its high catalytic activity and stability [27]. Various alternative CE materials such as carbon-based materials, metal oxides or chalcogenides, and alloys or intermetallics have been studied extensively [3,5,28–36]. Composites containing Pt and Sn have been used as electrocatalysts for CEs of DSSCs [36,37], methanol or ethanol oxidation [38–44], aqueous phase oxidation [45], and gas sensing [46]. The addition of metal oxides has been reported to improve the catalytic activity [40,47]. Pt:SnO$_2$ electrocatalytic films were used as CEs of DSSCs [48]. Dao et al. fabricated DSSCs with a PtSn alloy supported by reduced graphene oxides via dry plasma reduction [36]. In the present study, Pt-SnO$_x$ composites were synthesized by mixing chloroplatinic acid and tin(II) chloride that were processed using a DC-pulse nitrogen APPJ. X-ray photoelectron spectroscopy (XPS) results showed that the majority of Sn was in the oxidation state. The DSSC efficiency can be improved rapidly through 5 s APPJ processing of the chloroplatinic acid and tin(II) chloride mixture precursor; no metallic Pt was converted within such a short processing time. This suggests the catalytic effect of oxidized Pt and Sn compounds. A DSSC with a 120 s APPJ-processed Pt-SnO$_x$ CE shows efficiency comparable to that of a cell with a furnace-processed Pt CE.

2. Materials and Methods

2.1. Preparation of Pt-SnO$_x$ CEs

25-mM chloroplatinic acid (H$_2$PtCl$_6$) (purity: 99.95%, Uniregion Biotech, Taipei, Taiwan) and 25-mM tin(II) chloride (SnCl$_2$) isopropanol solutions were separately stirred for 24 h. These two solutions were mixed with the same volume ratios and were stirred using a magnetic stirrer (PC-420D, Corning Inc., Corning, NY, USA)for another 24 h. Next, 60 µL of the mixture precursor was spin-coated onto fluorine-doped tin oxide (FTO) substrates with an area of 1.5 cm × 1.5 cm at a speed of 1000 rpm for 15 s. The spin-coated precursors were then processed by a nitrogen APPJ for 5, 15, 30, 60, and 120 s. Figure 1a shows the APPJ setup. The operation parameters are as follows: nitrogen flow of 46 standard liter per minute (slm), power supply voltage of 275 V, and ON/OFF duty cycle of 7/33 µs. The temperature evolution of the substrates, shown in Figure 1b, was measured using a K-type thermocouple (OMEGA Engineering, Norwalk, CT, USA). The temperature rapidly increased to ~510 °C, and it dramatically decreased after the APPJ was turned off. Because our process is conducted at ~510 °C, we use FTO glass substrates (Sigma-Aldrich, St. Louis, MO, USA) which can tolerate a higher processing temperature.

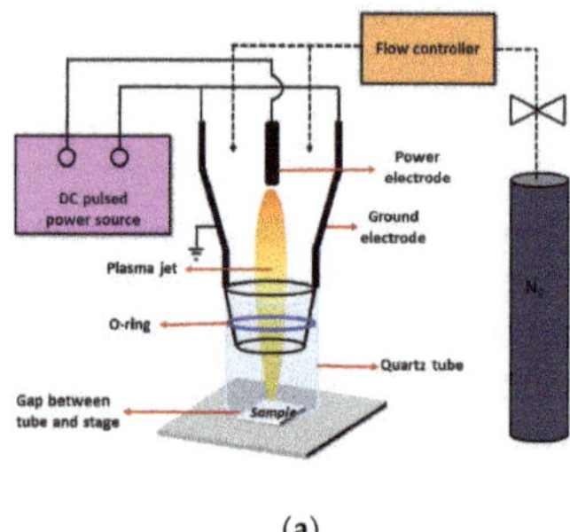

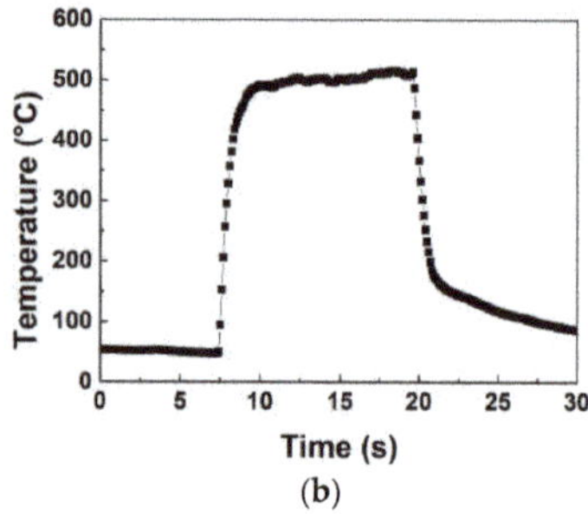

Figure 1. (**a**) Schematic of APPJ setup; (**b**) Temperature evolution of substrate during APPJ treatment.

2.2. Preparation of TiO$_2$ Photoanode and Assembly of DSSCs

The photoanode consists of a TiO$_2$ compact layer and a TiO$_2$ nanoporous layer for dye adsorption. First, a 0.23-M titanium isopropoxide solution (Fluka, St. Louis, MO, USA) was spin-coated on a FTO substrate and then baked at 200 °C for 10 min to form a TiO$_2$ compact layer to prevent electron recombination. Then, 1.6 g of TiO$_2$ nanoparticles (diameter: ~21 nm), 8 mL of ethanol, 6.49 g of terpineol (anhydrous, #86480, Fluka, St. Louis, MO, USA), 4.5 g of 10 wt % ethyl cellulose ethanolic solution (5–15 mPa·s, #46070, Fluka, St. Louis, MO, USA), and 3.5 g of 10 wt % ethyl cellulose ethanolic solution (30–50 mPa·s, #46080, Fluka, St. Louis, MO, USA) were mixed together. Next, a 0.4 g mixture containing TiO$_2$ was mixed with 500 μL of ethanol and stirred using a magnetic stirrer for 24 h. The mixed solution was baked at 53 °C until its weight became 0.175 g, thus completing the preparation of the TiO$_2$ pastes. The TiO$_2$ pastes were screen-printed onto the TiO$_2$ compact layer coated FTO substrate with a printed area of 0.5 cm × 0.5 cm. The screen-printed pastes were calcined at 510 °C for 15 min in a conventional furnace to form the TiO$_2$ photoanode. Next, the TiO$_2$ photoanode was immersed in a 0.3-mM N719 solution, which is mixed with acetonitrile and tertbutyl alcohol in a 1:1 volume ratio for 24 h. This completed the preparation of the dye-anchored nanoporous TiO$_2$ photoanodes.

The Pt-SnO$_x$ CEs and dye-anchored TiO$_2$ photoanodes were assembled with a 25-μm-thick spacer to form sandwich-structure DSSCs. Then, a commercial electrolyte (E-Solar EL 200, Everlight Chemical Industrial Co., Taipei, Taiwan) was injected into the solar cells.

Counterpart DSSC with furnace-processed Pt CE was fabricated for comparison. In this case, 60 μL of 25-mM H$_2$PtCl$_6$ isopropanol solution was spin-coated on the FTO substrate and calcined at 400 °C for 15 min using a tube furnace. The assembly procedure of DSSC with furnace-processed Pt CE is the same as that of DSSC with APPJ-processed Pt-SnO$_x$ CE.

2.3. Characterization of Materials and DSSCs

During the APPJ reduction processes, a spectrometer (USB4000, Ocean Optics, Largo, FL, USA) was used for monitoring the plasma optical emission spectra (OES). Pt-SnO$_x$ nanoparticles were inspected using a scanning electron microscope (SEM, JSM-7800F Prime, JEOL, Tokyo, Japan) with an energy-dispersive spectroscopy (EDS) attachment. To investigate the chemical configuration of Pt-SnO$_x$, XPS (Thermo K-Alpha, VGS, Waltham, MA, USA was used for analyzing the binding status. The C1s core level was centered at 284.6 eV to calibrate the binding energy scale. XPSPEAK 4.1 software (was used for fitting binding energy positions. XPS samples were prepared with Corning glass substrates instead of FTO glass ones to avoid the interference of Sn signals emitted from FTO substrates. To examine the electrochemical catalytic activities of Pt-SnO$_x$ CEs, electrochemical impedance spectroscopy (EIS) and Tafel measurements were performed using an electrochemical workstation (PGSTAT204, Metrohm Autolab, Herisau, Switzerland). EIS measurements were performed with a sinusoidal amplitude of 10 mV with frequencies of 0.1-10^5 Hz, and the data were fitted using Z-view 3.1 software. Tafel curves were recorded from −0.6 V to 0.6 V at a scan rate of 50 mV/s. Both measurements were performed on a symmetrical cell with two equal Pt-SnO$_x$ CEs. A solar simulator (WXS-155S-L2, WACOM, Saitama, Japan) with an AM 1.5 filter equipped with an electrometer (Keithley 2440, Tektronix, Beaverton, OR, USA) was used for measuring the photocurrent-voltage characteristics of the DSSCs.

3. Results and Discussion

Figure 2a shows the plasma OES evolution during APPJ processing of the mixed H$_2$PtCl$_6$/SnCl$_2$ precursor. NO$_\gamma$, NO$_\beta$, N$_2$ 1st positive, and N$_2$ 2nd positive emissions were observed clearly during 120 s APPJ processes. Figure 2b shows the plasma spectra when processing H$_2$PtCl$_6$, SnCl$_2$, and mixed H$_2$PtCl$_6$/SnCl$_2$ precursors on the FTO substrates. The NO$_\gamma$ system (A$^2\Sigma^+$-X$^2\Pi$) is located at wavelengths lower than 280 nm. The NO$_\beta$ system (B$^2\Pi$-X$^2\Pi$) is located from around 260 to 500 nm, and it partially overlaps the NO_γ system. The other emissions at 357, 385, and 389 nm were attributed to the N_2 2nd positive system ($C^3\Pi_u$-$B^3\Pi_g$); these overlap with the NO_β system. The N_2 1st positive system ($B^3\Pi_g$-$A^3\Sigma_u{}^+$) was located at wavelengths higher than 530 nm.

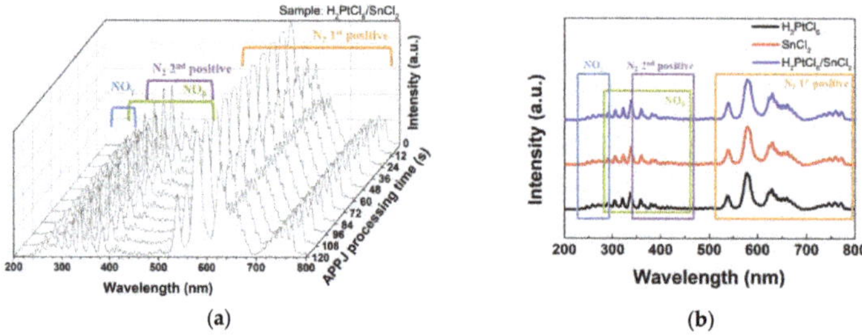

Figure 2. (a) OES evolution during APPJ processing of mixed H_2PtCl_6/$SnCl_2$ precursors. (b) OES when processing H_2PtCl_6, $SnCl_2$, and mixed H_2PtCl_6/$SnCl_2$ precursors using nitrogen APPJ.

Figure 3a–e shows the SEM images of Pt-SnO$_x$ nanoparticles converted from mixed H_2PtCl_6/$SnCl_2$ precursors on the FTO glass substrates using various APPJ processing times. The nanoparticle size and morphology remained similar for APPJ processing times of 5-120 s. Figure 3f shows EDS results for the 120 s and APPJ-processed sample. Pt and Sn signals indicate the presence of two elements in the nanoparticles. Both of Sn and O signals could result from the nanoparticles and the FTO substrates.

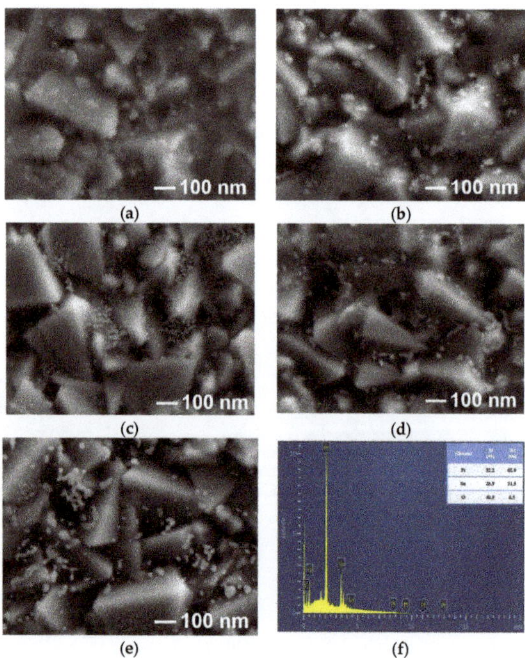

Figure 3. Scanning electron microscope (SEM) images of samples processed by APPJ for various durations: (a) 5 s, (b) 15 s, (c) 30 s, (d) 60 s, and (e) 120 s. (f) Energy-dispersive spectroscopy (EDS) spectrum of nanoparticles converted from H_2PtCl_6/$SnCl_2$ precursors on FTO glass substrates using 120 s APPJ processing.

To identify the chemical states of Pt-SnO$_x$ compounds, Figure 4a,b shows the XPS spectra of Pt4f and Sn3d for samples. The Pt4f spectrum can be deconvoluted into three components including Pt, Pt^{2+}, and Pt^{4+}. The metallic peaks of Pt are located at 71.30 and 74.65 eV, Pt(II) components are located at 72.70 and 76.50 eV, and Pt(IV) components are located at 73.80 and 77.15 eV [49,50]. In Figure 4a, the major peaks belong to Pt^{2+} and Pt^{4+} for as-deposited and 5 and 15 s APPJ-processed samples. These results indicate that most of the H$_2$PtCl$_6$/SnCl$_2$ precursor was not converted to metallic Pt by APPJ processing for less than 15 s. As the APPJ processing time increases, increased conversion of precursors into metallic Pt was clearly observed. The Pt^{2+} signal is noted as the oxidation state of Pt, and it could indicate PtO [51,52] or Pt(OH)$_2$ [53]. The presence of Pt oxidation states, due to the interaction with the Pt-support, is attributed to an electronic effect or oxygen absorption from air [54,55]. Figure 4b shows the oxidation state of Sn3d under various APPJ processing times. The binding energy of Sn3d can be deconvoluted into two categories: one at 485.8 and 494.2 eV for the zero-valent state of Sn, and the other at 487.3 and 495.7 eV for Sn(II/IV) components [56]. The major peak is attributed to the oxidation state of Sn for up to 120 s, and the percentage of metallic Sn increased only slightly increased with the APPJ processing time. Sn(II) and Sn(IV) species are difficult to distinguish from XPS measurements because of the small difference between their binding energies [57,58]. Tables 1 and 2 show the percentages of Pt and Sn species, respectively. The Pt-support interaction may influence charge transfer from Pt to oxygen species on the surface and improve the electrochemical catalytic abilities and catalyst stability [47].

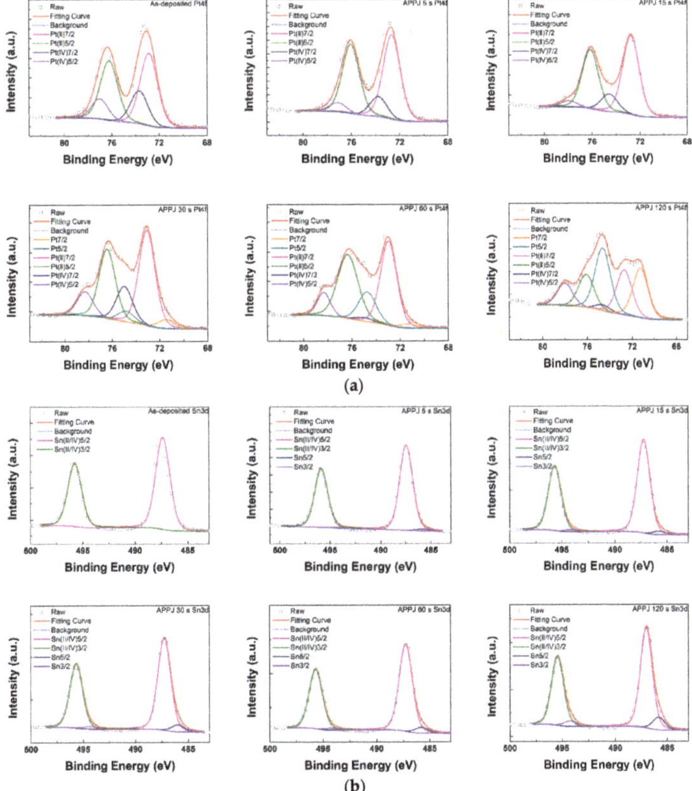

Figure 4. X-ray photoelectron spectroscopy (XPS) spectra of (**a**) Pt4f and (**b**) Sn3d for samples processed by APPJ for various durations.

Table 1. Percentage of Pt species obtained from XPS analysis.

APPJ Pt4f (%)	Pt 7/2	Pt 5/2	Pt(II) 7/2	Pt(II) 5/2	Pt(IV) 7/2	Pt(IV) 5/2
0 s	-	-	39.19	32.34	17.90	10.58
5 s	-	-	46.17	37.94	10.90	4.99
15 s	-	-	48.87	36.29	10.87	3.97
30 s	3.72	4.38	39.16	28.52	14.39	9.82
60 s	1.87	14.34	38.04	35.64	1.80	8.31
120 s	22.66	28.72	20.46	15.21	2.69	10.26

Table 2. Percentage of Sn species obtained from XPS analysis.

APPJ Sn3d (%)	Sn 5/2	Sn 3/2	Sn(II/IV) 5/2	Sn(II/IV) 3/2
0 s	-	-	60.19	39.81
5 s	1.13	0.36	58.33	40.19
15 s	2.21	1.31	56.95	39.53
30 s	4.17	1.72	55.35	38.76
60 s	3.72	1.14	56.27	38.87
120 s	6.72	2.93	53.51	36.84

Figure 5a,b shows the EIS Nyquist and Bode phase plots to evaluate the catalytic activities of APPJ-processed Pt-SnO_x CEs. The inset of Figure 5a shows the equivalent circuit for Nyquist curve fitting [59]. The series resistance (R_s) and charge-transfer resistance (R_{ct}) can be described as the resistance of substrates and the catalytic effect of the electrode-reducing triiodide ions, respectively. R_s can be obtained from the high-frequency intercept on the real axis and R_{ct}, from the radius of the real semi-circle [60]. Table 3 shows the EIS parameters including R_s, R_{ct}, and constant phase element (CPE1) [29]. A higher catalytic effect and lower charge-transfer resistance would improve the DSSC performance. For all cases, R_s of Pt-SnO_x CEs remained similar. R_{ct} generally decreased (i.e., semi-circle became smaller) as the APPJ processing time increased, indicating that APPJ processing can enhance the catalytic activity. R_{ct} was comparable for APPJ processing times of 60 s (4.72 Ω) and 120 s (4.69 Ω). Lower R_{ct} results in a higher electrocatalytic activity at the interface between the CEs and the electrolytes [61]. CPE1, which represents the interfacial capacitance between the electrode and the electrolyte, is also a good indicator of the surface activity of CEs [62–64]. The 120 s APPJ-processed CEs had a higher CPE1-T (105.5 $\mu F/cm^2$), indicating larger surface reaction between the CE and the electrolyte. Bode phase plots show the electron lifetime for recombination in devices; the electron lifetime is expressed as $\tau_e = 1/(2\pi f_{peak})$, where f_{peak} is the frequency of the highest peak. Shorter electron lifetime indicates faster charge transfer at the interface between the CE and the electrolyte [64,65]. In Figure 5b, the trend of the electron lifetime follows the EIS results. The 5 s APPJ-processed CE has the smallest peak frequency, indicating the largest electron lifetime with slower charge transfer. Furthermore, electron lifetimes are comparable in 60 s and 120 s APPJ-processed CEs, and this is consistent with the results for R_{ct}.

To further clarify the catalytic activities of Pt-SnO_x CEs, Tafel polarization experiments were conducted and the results are shown in Figure 6. The exchange current density (J_0) was measured by the intercept of the Y-axis (zero voltage) from the tangential line of the curve [66,67]. The 120 s APPJ-processed CEs had a large J_0, indicating better electrocatalyic activity and lower charge-transfer resistance at the interface of the CE and the electrolyte. Table 3 shows that J_0 increases with the APPJ processing time. APPJ processes enhanced the triiodide reduction reaction [60]. The exchange current density is also proportional to R_{ct} obtained from the EIS measurement. It can be described as $J_0 = RT/nFR_{ct}$, where R is a gas constant; T is temperature; n is the number of electrons involved in the redox reaction; and F is the Faraday's constant [68]. EIS and Tafel measurements both indicate that APPJ-processed Pt-SnO_x elecrodes show suitable catalytic performance for use as the CEs of DSSCs.

J_0 increases with the APPJ processing time, indicating that APPJ processing can enhance the catalyst activity of Pt-SnO$_x$.

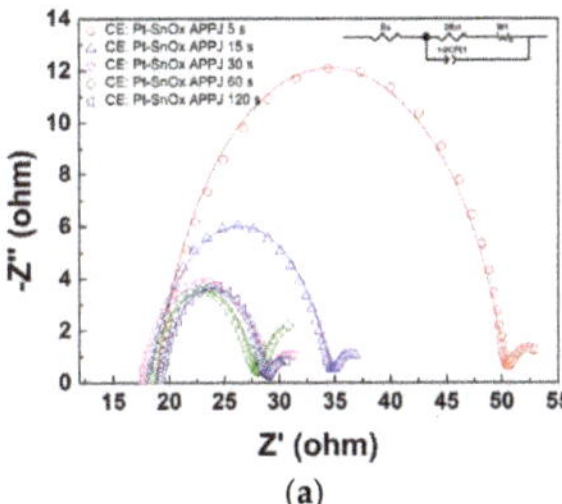

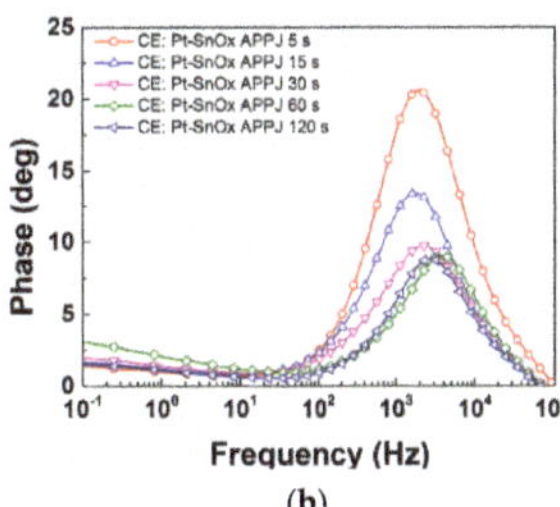

Figure 5. (**a**) Nyquist curves of symmetric cells with two Pt-SnO$_x$ CEs. The inset shows the equivalent circuit diagram. (**b**) Bode phase plots of symmetric cells with two Pt-SnO$_x$ CEs.

Table 3. EIS parameters of Pt-SnO$_x$ CEs.

Counter Electrode		R_s (Ω)	R_{ct} (Ω)	CPE1-T (μF/cm^2)	CPE1-P	W1			J_0 [a] (mA/cm^2)	J_0 [b] (mA/cm^2)
						W1-R(Ω)	W1-T(s)	W1-P		
	5 s	18.8	15.74	35.6	0.835	3.23	2.13	0.5	0.82	1.04
	15 s	18.2	8.18	83.1	0.808	2.82	1.99	0.5	1.58	1.39
Pt-SnO$_x$ APPJ	30 s	17.55	5.8	95	0.75	2.73	2.25	0.5	2.23	1.67
	60 s	18.25	4.72	105	0.815	5.04	3.15	0.5	2.74	1.85
	120 s	19.28	4.69	105.5	0.84	2.21	1.8	0.5	2.75	2.08

[a] J_0: Exchange current density is calculated from R_{ct}. [b] J_0: Exchange current density is calculated from Tafel curve.

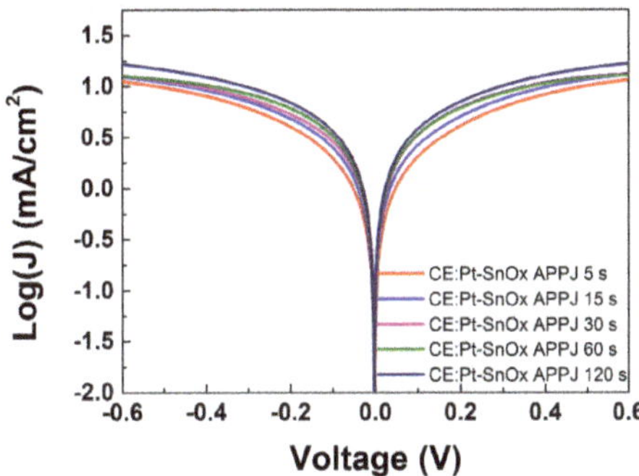

Figure 6. Tafel curves of symmetric cells with various Pt-SnO$_x$ CEs.

Figure 7 shows the IV curves of DSSCs with APPJ-processed Pt-SnO$_x$ CEs. Table 4 shows the photovoltaic parameters, including the open-circuit voltage (V_{oc}), short-circuit current (J_{sc}), fill factor (FF), and efficiency (EFF) with their standard deviations. The power conversion efficiencies (PCEs) of DSSCs with 5 s and 15 s APPJ-processed Pt-SnO$_x$ CEs are 3.87 $\pm$ 0.58% and 3.86 $\pm$ 0.28%, respectively, indicating that APPJ processing for a short duration can improve the DSSC performance. XPS results show that almost no metallic Pt was converted with 5 s and 15 s APPJ processing, indicating the catalytic effect of oxidized Pt and Sn compound CEs in DSSCs, and this agrees with previous reported findings [30,32]. As the APPJ treatment time increases, the PCE of DSSCs with 30 s, 60 s, and 120 s APPJ-processed CEs reaches 4.01 $\pm$ 0.34%, 4.20 $\pm$ 0.41%, and 4.46 $\pm$ 0.29%, respectively. The performance of DSSC with a 120 s APPJ-processed Pt-SnO$_x$ CE was comparable to that with a conventional furnace-processed Pt CE (4.42 ± 0.26%). Figure 8 shows the statistics of the DSSC parameters. APPJ processing gradually increased the FFs and PCEs of DSSCs, consistent with the results obtained from EIS and Tafel measurement. The improved FF and efficiency with APPJ processing time could result from the better conversion of metallic Pt from the precursor solution.

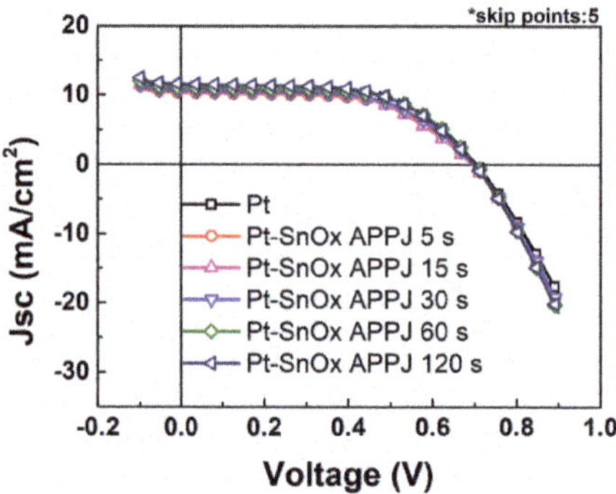

Figure 7. Photocurrent density-voltage curves of DSSCs with various CEs.

Table 4. Photovoltaic parameters of DSSCs with different CEs.

Condition		V_{oc} (V)	J_{sc} (mA/cm^2)	FF (%)	EFF (%)
Pt		0.70 ± 0.02	10.34 ± 0.47	61.27 ± 1.91	4.42 ± 0.26
	5 s	0.70 ± 0.01	10.76 ± 0.82	51.65 ± 6.03	3.87 ± 0.58
	15 s	0.69 ± 0.02	10.30 ± 0.78	54.43 ± 2.65	3.86 ± 0.28
Pt-SnO$_x$ APPJ	30 s	0.69 ± 0.02	10.26 ± 0.80	56.38 ± 2.37	4.01 ± 0.34
	60 s	0.69 ± 0.02	10.47 ± 0.94	57.91 ± 2.04	4.20 ± 0.41
	120 s	0.70 ± 0.02	10.72 ± 0.75	59.41 ± 1.08	4.46 ± 0.29

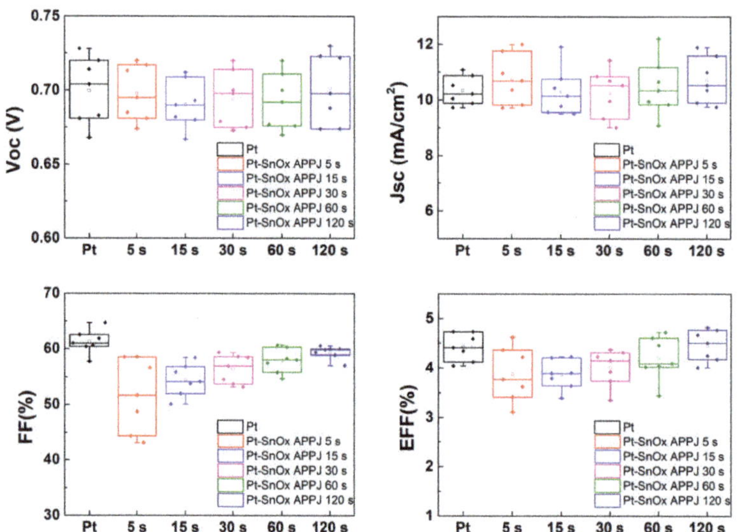

Figure 8. Statistics of DSSC parameters based on APPJ-processed Pt-SnO$_x$ CEs and furnace-processed Pt CE (reference).

4. Conclusions

We analyze the time evolution of Pt-SnO$_x$ nanoparticle catalysts that are converted from a mixture of chloroplatinic acid and tin(II) chloride using DC-pulse nitrogen APPJ. XPS analyses indicate the conversion of a large portion of the metallic Pt and tin oxide. EIS and Tafel measurements indicate improved electrochemical catalytic effects. The synthesized Pt-SnO$_x$ nanoparticles on FTO glass substrates are used as the CEs of DSSCs. The I-V curve shows that the performance of DSSCs with APPJ-processed Pt-SnO$_x$ CEs is comparable to that of DSSCs with conventional furnace-processed Pt CEs. As the APPJ processing time is increased, the FF and efficiency of DSSCs gradually increase. Our results show that a DC-pulse nitrogen APPJ is an efficient tool for synthesizing Pt-SnO$_x$ catalysts from a mixture precursor solution consisting of chloroplatinic acid and tin(II) chloride.

Author Contributions: C.-C.L. performed the experiments, analyzed the data, and wrote the paper draft. T.-M.H. assisted in conducting experiments. I.-C.C. and C.-C.H. assisted in instructing the research and analyzing the data. J.-Z.C. directed the research direction, analyzed the data, and revised the paper. All authors commented on the manuscript.

Funding: This work is supported by the "Advanced Research Center for Green Materials Science and Technology" from The Featured Area Research Center Program of the Higher Education Sprout Project by the Ministry of Education (107L9006) and the Ministry of Science and Technology in Taiwan (MOST 105-2221-E-002-047-MY3, MOST 106-2221-E-002-193-MY2 & MOST 107-3017-F-002-001).

Acknowledgments: The cleanroom facility is provided by the Nano-Electro-Mechanical-Systems (NEMS) Research Center at National Taiwan University. Yuan-Tzu Lee of the Instrumentation Center at National Taiwan University helps with the SEM operation.

References

1. Schutze, A.; Jeong, J.Y.; Babayan, S.E.; Park, J.; Selwyn, G.S.; Hicks, R.F. The atmospheric-pressure plasma jet: A review and comparison to other plasma sources. *IEEE Trans. Plasma Sci.* **1998**, *26*, 1685–1694. [CrossRef]

2. Laroussi, M.; Akan, T. Arc-free atmospheric pressure cold plasma jets: A review. *Plasma Proc. Polym.* **2007**, *4*, 777–788. [CrossRef]

3. Chen, J.-Z.; Wang, C.; Hsu, C.-C.; Cheng, I.-C. Ultrafast synthesis of carbon-nanotube counter electrodes for dye-sensitized solar cells using an atmospheric-pressure plasma jet. *Carbon* **2016**, *98*, 34–40. [CrossRef]

4. Chou, C.-Y.; Chang, H.; Liu, H.-W.; Yang, Y.-J.; Hsu, C.-C.; Cheng, I.-C.; Chen, J.-Z. Atmospheric-pressure-plasma-jet processed nanoporous TiO$_2$ photoanodes and Pt counter-electrodes for dye-sensitized solar cells. *RSC Adv.* **2015**, *5*, 45662–45667. [CrossRef]

5. Wan, T.-H.; Chiu, Y.-F.; Chen, C.-W.; Hsu, C.-C.; Cheng, I.; Chen, J.-Z. Atmospheric-pressure plasma jet processed Pt-decorated reduced graphene oxides for counter-electrodes of dye-sensitized solar cells. *Coatings* **2016**, *6*, 44. [CrossRef]

6. Wu, T.J.; Chou, C.Y.; Hsu, C.M.; Hsu, C.C.; Chen, J.Z.; Cheng, I.C. Ultrafast synthesis of continuous Au thin films from chloroauric acid solution using an atmospheric pressure plasma jet. *RSC Adv.* **2015**, *5*, 99654–99657. [CrossRef]

7. Liu, L.; Ye, D.; Yu, Y.; Liu, L.; Wu, Y. Carbon-based flexible micro-supercapacitor fabrication via mask-free ambient micro-plasma-jet etching. *Carbon* **2017**, *111*, 121–127. [CrossRef]

8. Kuok, F.-H.; Kan, K.-Y.; Yu, I.-S.; Chen, C.-W.; Hsu, C.-C.; Cheng, I.-C. Application of atmospheric-pressure plasma jet processed carbon nanotubes to liquid and quasi-solid-state electrolyte supercapacitors. *Appl. Surf. Sci.* **2017**, *425*, 321–328. [CrossRef]

9. Yang, C.-H.; Kuok, F.-H.; Liao, C.-Y.; Wan, T.-H.; Chen, C.-W.; Hsu, C.-C.; Cheng, I.-C.; Chen, J.-Z. Flexible reduced graphene oxide supercapacitor fabricated using a nitrogen dc-pulse atmospheric-pressure plasma jet. *Mater. Res. Express* **2017**, *4*, 025504. [CrossRef]

10. Liu, H.W.; Liang, S.P.; Wu, T.J.; Chang, H.; Kao, P.K.; Hsu, C.C.; Chen, J.Z.; Chou, P.T.; Cheng, I.C. Rapid atmospheric pressure plasma jet processed reduced graphene oxide counter electrodes for dye-sensitized solar cells. *ACS Appl. Mater. Interfaces* **2014**, *6*, 15105–15112. [CrossRef] [PubMed]

11. Kuok, F.H.; Liao, C.Y.; Wan, T.H.; Yeh, P.W.; Cheng, I.C.; Chen, J.Z. Atmospheric pressure plasma jet processed reduced graphene oxides for supercapacitor application. *J. Alloys Compd.* **2017**, *692*, 558–562. [CrossRef]

12. Chen, S.L.; Wang, S.; Wang, Y.B.; Guo, B.H.; Li, G.Q.; Chang, Z.S.; Zhang, G.J. Surface modification of epoxy resin using He/CF_4 atmospheric pressure plasma jet for flashover withstanding characteristics improvement in vacuum. *Appl. Surf. Sci.* **2017**, *414*, 107–113. [CrossRef]

13. Munoz, J.; Bravo, J.A.; Calzada, M.D. Aluminum metal surface cleaning and activation by atmospheric-pressure remote plasma. *Appl. Surf. Sci.* **2017**, *407*, 72–81. [CrossRef]

14. Tsai, J.-H.; Cheng, I.-C.; Hsu, C.-C.; Chen, J.-Z. Dc-pulse atmospheric-pressure plasma jet and dielectric barrier discharge surface treatments on fluorine-doped tin oxide for perovskite solar cell application. *J. Phys. D: Appl. Phys.* **2017**, *51*, 025502. [CrossRef]

15. Bose, A.C.; Shimizu, Y.; Mariotti, D.; Sasaki, T.; Terashima, K.; Koshizaki, N. Flow rate effect on the structure and morphology of molybdenum oxide nanoparticles deposited by atmospheric-pressure microplasma processing. *Nanotechnology* **2006**, *17*, 5976–5982. [CrossRef]

16. Babayan, S.; Jeong, J.; Schütze, A.; Tu, V.; Moravej, M.; Selwyn, G.; Hicks, R. Deposition of silicon dioxide films with a non-equilibrium atmospheric-pressure plasma jet. *Plasma Sources Sci. Technol.* **2001**, *10*, 573. [CrossRef]

17. Patel, J.; Nemcova, L.; Maguire, P.; Graham, W.G.; Mariotti, D. Synthesis of surfactant-free electrostatically stabilized gold nanoparticles by plasma-induced liquid chemistry. *Nanotechnology* **2013**, *24*, 245604. [CrossRef] [PubMed]

18. Lee, C.-C.; Wan, T.-H.; Hsu, C.-C.; Cheng, I.-C.; Chen, J.-Z. Atmospheric-pressure plasma jet processed Pt/ZnO composites and its application as counter-electrodes for dye-sensitized solar cells. *Appl. Surf. Sci.* **2018**, *436*, 690–696. [CrossRef]

19. Malecki, S.; Gargul, K. Low-waste recycling of spent CuO-ZnO-Al_2O_3 catalysts. *Metals* **2018**, *8*, 177. [CrossRef]

20. Paiva, A.P. Recycling of palladium from spent catalysts using solvent extraction-some critical points. *Metals* **2017**, *7*, 505. [CrossRef]

21. Fujita, T.; Higuchi, K.; Yamamoto, Y.; Tokunaga, T.; Arai, S.; Abe, H. In-situ TEM study of a nanoporous Ni-Co catalyst used for the dry reforming of methane. *Metals* **2017**, *7*, 406. [CrossRef]

22. Joo, S.H.; Shin, D.J.; Oh, C.H.; Wang, J.P.; Park, J.T.; Shin, S.M. Application of Co and Mn for a Co-Mn-Br or Co-Mn-$C_2H_3O_2$ petroleum liquid catalyst from the cathode material of spent lithium ion batteries by a hydrometallurgical route. *Metals* **2017**, *7*, 439. [CrossRef]

23. Tai, M.C.; Gentle, A.; de Silva, K.S.B.; Arnold, M.D.; van der Lingen, E.; Cortie, M.B. Thermal stability of nanoporous raney gold catalyst. *Metals* **2015**, *5*, 1197–1211. [CrossRef]

24. Sabitu, S.T.; Goudy, A.J. Dehydrogenation kinetics and modeling studies of MgH_2 enhanced by transition metal oxide catalysts using constant pressure thermodynamic driving forces. *Metals* **2012**, *2*, 219–228. [CrossRef]

25. Song, H.; Hu, F.Y.; Peng, Y.; Li, K.Z.; Bai, S.P.; Li, J.H. Non-thermal plasma catalysis for chlorobenzene removal over CoMn/TiO_2 and CeMn/TiO_2: Synergistic effect of chemical catalysis and dielectric constant. *Chem. Eng. J.* **2018**, *347*, 447–454. [CrossRef]

26. O'regan, B.; Grfitzeli, M. A low-cost, high-efficiency solar cell based on dye-sensitized. *Nature* **1991**, *353*, 737–740. [CrossRef]

27. Lin, C.-Y.; Lin, J.-Y.; Wan, C.-C.; Wei, T.-C. High-performance and low platinum loading electrodeposited-Pt counter electrodes for dye-sensitized solar cells. *Electrochim. Acta* **2011**, *56*, 1941–1946. [CrossRef]

28. Bae, K.-H.; Park, E.; Dao, V.-D.; Choi, H.-S. PtZn nanoalloy counter electrodes as a new avenue for highly efficient dye-sensitized solar cells. *J. Alloys Compd.* **2017**, *702*, 449–457. [CrossRef]

29. Dao, V.D.; Nang, L.V.; Kim, E.T.; Lee, J.K.; Choi, H.S. Pt nanoparticles immobilized on CVD-grown graphene as a transparent counter electrode material for dye-sensitized solar cells. *ChemSusChem* **2013**, *6*, 1316–1319. [CrossRef] [PubMed]

30. Li, C.-T.; Chang, H.-Y.; Li, Y.-Y.; Huang, Y.-J.; Tsai, Y.-L.; Vittal, R.; Sheng, Y.-J.; Ho, K.-C. Electrocatalytic zinc composites as the efficient counter electrodes of dye-sensitized solar cells: Study on the electrochemical performances and density functional theory calculations. *ACS Appl. Mater. Interfaces* **2015**, *7*, 28254–28263. [CrossRef] [PubMed]

31. Wan, J.; Fang, G.; Yin, H.; Liu, X.; Liu, D.; Zhao, M.; Ke, W.; Tao, H.; Tang, Z. Pt–Ni alloy nanoparticles as superior counter electrodes for dye-sensitized solar cells: Experimental and theoretical understanding. *Adv. Mater.* **2014**, *26*, 8101–8106. [CrossRef] [PubMed]

32. Wang, H.; Wei, W.; Hu, Y.H. Efficient ZnO-based counter electrodes for dye-sensitized solar cells. *J. Mater. Chem. A* **2013**, *1*, 6622–6628. [CrossRef]

33. Yang, Q.; Duan, J.; Yang, P.; Tang, Q. Counter electrodes from platinum alloy nanotube arrays with zno nanorod templates for dye-sensitized solar cells. *Electrochim. Acta* **2016**, *190*, 648–654. [CrossRef]

34. Tang, Q.; Duan, J.; Duan, Y.; He, B.; Yu, L. Recent advances in alloy counter electrodes for dye-sensitized solar cells. A critical review. *Electrochim. Acta* **2015**, *178*, 886–899. [CrossRef]

35. Nechiyil, D.; Vinayan, B.P.; Ramaprabhu, S. Tri-iodide reduction activity of ultra-small size ptfe nanoparticles supported nitrogen-doped graphene as counter electrode for dye-sensitized solar cell. *J. Colloid Interface Sci.* **2017**, *488*, 309–316. [CrossRef] [PubMed]

36. Jin, I.-K.; Dao, V.-D.; Larina, L.L.; Choi, H.-S. Optimum engineering of a PtSn alloys/reduced graphene oxide nanohybrid for a highly efficient counter electrode in dye-sensitized solar cells. *J. Ind. Eng. Chem.* **2016**, *36*, 238–244. [CrossRef]

37. Zhu, Y.J.; Gao, C.J.; Han, Q.J.; Wang, Z.; Wang, Y.R.; Zheng, H.K.; Wu, M.X. Large-scale high-efficiency dye-sensitized solar cells based on a Pt/carbon spheres composite catalyst as a flexible counter electrode. *J. Catal.* **2017**, *346*, 62–69. [CrossRef]

38. Jiang, L.; Sun, G.; Sun, S.; Liu, J.; Tang, S.; Li, H.; Zhou, B.; Xin, Q. Structure and chemical composition of supported Pt–Sn electrocatalysts for ethanol oxidation. *Electrochim. Acta* **2005**, *50*, 5384–5389. [CrossRef]

39. Baranova, E.A.; Padilla, M.A.; Halevi, B.; Amir, T.; Artyushkova, K.; Atanassov, P. Electrooxidation of ethanol on PtSn nanoparticles in alkaline solution: Correlation between structure and catalytic properties. *Electrochim. Acta* **2012**, *80*, 377–382. [CrossRef]

40. Pang, H.; Lu, J.; Chen, J.; Huang, C.; Liu, B.; Zhang, X. Preparation of SnO_2-CNTs supported Pt catalysts and their electrocatalytic properties for ethanol oxidation. *Electrochim. Acta* **2009**, *54*, 2610–2615. [CrossRef]

41. Ordonez, L.C.; Roquero, P.; Sebastian, P.J.; Ramirez, J. Carbon-supported platinum-molybdenum electro-catlysts for methanol oxidation. *Catal. Today* **2005**, *107*, 46–52. [CrossRef]

42. Jiang, L.H.; Sun, G.Q.; Zhou, Z.H.; Xin, Q. Preparation and characterization of PtSn/C anode electrocatalysts for direct ethanol fuel cell. *Catal. Today* **2004**, *93*, 665–670. [CrossRef]

43. Jiang, W.; Pang, Y.J.; Gu, L.L.; Yao, Y.; Su, Q.; Ji, W.J.; Au, C.T. Structurally defined SnO_2 substrates, nanostructured Au/SnO_2 interfaces, and their distinctive behavior in benzene and methanol oxidation. *J. Catal.* **2017**, *349*, 183–196. [CrossRef]

44. Lee, B.; Sakamoto, Y.; Hirabayashi, D.; Suzuki, K.; Hibino, T. Direct oxidation of methane to methanol over proton conductor/metal mixed catalysts. *J. Catal.* **2010**, *271*, 195–200. [CrossRef]

45. Xie, J.H.; Falcone, D.D.; Davis, R.J. Restructuring of supported PtSn bimetallic catalysts during aqueous phase oxidation of 1,6-hexanediol. *J. Catal.* **2015**, *332*, 38–50. [CrossRef]

46. Mädler, L.; Roessler, A.; Pratsinis, S.E.; Sahm, T.; Gurlo, A.; Barsan, N.; Weimar, U. Direct formation of highly porous gas-sensing films by in situ thermophoretic deposition of flame-made Pt/SnO_2 nanoparticles. *Sens. Actuators B Chem.* **2006**, *114*, 283–295. [CrossRef]

47. Antolini, E. Formation of carbon-supported PtM alloys for low temperature fuel cells: A review. *Mater. Chem. Phys.* **2003**, *78*, 563–573. [CrossRef]

48. Khelashvili, G.; Behrens, S.; Hinsch, A.; Habicht, W.; Schild, D.; Eichhöfer, A.; Sastrawan, R.; Skupien, K.; Dinjus, E.; Bönnemann, H. Preparation and characterization of low platinum loaded Pt: SnO_2 electrocatalytic films for screen printed dye solar cell counter electrode. *Thin Solid Films* **2007**, *515*, 4074–4079. [CrossRef]

49. Sun, H.; Ullah, R.; Chong, S.; Ang, H.M.; Tadé, M.O.; Wang, S. Room-light-induced indoor air purification using an efficient Pt/N-TiO_2 photocatalyst. *Appl. Catal. B Environ.* **2011**, *108*, 127–133. [CrossRef]

50. Ma, Y.; Wang, H.; Ji, S.; Linkov, V.; Wang, R. PtSn/C catalysts for ethanol oxidation: The effect of stabilizers on the morphology and particle distribution. *J. Power Sources* **2014**, *247*, 142–150. [CrossRef]

51. Xu, J.; Liu, X.; Chen, Y.; Zhou, Y.; Lu, T.; Tang, Y. Platinum–cobalt alloy networks for methanol oxidation electrocatalysis. *J. Mater. Chem.* **2012**, *22*, 23659–23667. [CrossRef]

52. Zhang, Y.; Han, T.; Zhu, L.; Fang, J.; Xu, J.; Xu, P.; Li, X.; Liu, C.-C. Pt35Cu65 nanoarchitecture: A highly durable and effective electrocatalyst towards methanol oxidation. *Nanotechnology* **2015**, *26*, 135706. [CrossRef] [PubMed]

53. Silva, J.; De Souza, R.F.; Romano, M.A.; D'Villa-Silva, M.; Calegaro, M.L.; Hammer, P.; Neto, A.O.; Santos, M.C. PtSnIr/C anode electrocatalysts: Promoting effect in direct ethanol fuel cells. *J. Braz. Chem. Soc.* **2012**, *23*, 1146–1153. [CrossRef]

54. García-Rodríguez, S.; Somodi, F.; Borbáth, I.; Margitfalvi, J.L.; Peña, M.A.; Fierro, J.L.G.; Rojas, S. Controlled synthesis of Pt-Sn/C fuel cell catalysts with exclusive Sn–Pt interaction: Application in co and ethanol electrooxidation reactions. *Appl. Catal. B Environ.* **2009**, *91*, 83–91. [CrossRef]

55. De la Fuente, J.G.; Rojas, S.; Martínez-Huerta, M.; Terreros, P.; Pena, M.; Fierro, J. Functionalization of carbon support and its influence on the electrocatalytic behaviour of Pt/C in H_2 and Co electrooxidation. *Carbon* **2006**, *44*, 1919–1929. [CrossRef]

56. Vilella, I.; De Miguel, S.; Scelza, O. Hydrogenation of citral on Pt and PtSn supported on activated carbon felts (ACF). *Latin Am. Appl. Res.* **2005**, *35*, 51–57.

57. Vilella, I.M.; de Miguel, S.R.; de Lecea, C.S.-M.; Linares-Solano, Á.; Scelza, O.A. Catalytic performance in citral hydrogenation and characterization of PtSn catalysts supported on activated carbon felt and powder. *Appl. Catal. A Gen.* **2005**, *281*, 247–258. [CrossRef]

58. Crabb, E.M.; Marshall, R.; Thompsett, D. Carbon monoxide electro-oxidation properties of carbon-supported PtSn catalysts prepared using surface organometallic chemistry. *J. Electrochem. Soc.* **2000**, *147*, 4440–4447. [CrossRef]

59. Gong, F.; Wang, H.; Wang, Z.-S. Self-assembled monolayer of graphene/Pt as counter electrode for efficient dye-sensitized solar cell. *Phys. Chem. Chem. Phys.* **2011**, *13*, 17676–17682. [CrossRef] [PubMed]

60. Yue, G.; Wu, J.; Xiao, Y.; Lin, J.; Huang, M.; Lan, Z.; Fan, L. Functionalized graphene/poly (3, 4-ethylenedioxythiophene): Polystyrenesulfonate as counter electrode catalyst for dye-sensitized solar cells. *Energy* **2013**, *54*, 315–321. [CrossRef]

61. Yin, X.; Xue, Z.; Liu, B. Electrophoretic deposition of Pt nanoparticles on plastic substrates as counter electrode for flexible dye-sensitized solar cells. *J. Power Sources* **2011**, *196*, 2422–2426. [CrossRef]

62. Sigdel, S.; Dubey, A.; Elbohy, H.; Aboagye, A.; Galipeau, D.; Zhang, L.; Fong, H.; Qiao, Q. Dye-sensitized solar cells based on spray-coated carbon nanofiber/TiO_2 nanoparticle composite counter electrodes. *J. Mater. Chem. A* **2014**, *2*, 11448–11453. [CrossRef]

63. Wang, Q.; Moser, J.-E.; Grätzel, M. Electrochemical impedance spectroscopic analysis of dye-sensitized solar cells. *J. Phys. Chem. B* **2005**, *109*, 14945–14953. [CrossRef] [PubMed]

64. Agarwala, S.; Thummalakunta, L.; Cook, C.; Peh, C.; Wong, A.; Ke, L.; Ho, G. Co-existence of LiI and KI in filler-free, quasi-solid-state electrolyte for efficient and stable dye-sensitized solar cell. *J. Power Sources* **2011**, *196*, 1651–1656. [CrossRef]

65. Jeong, H.; Pak, Y.; Hwang, Y.; Song, H.; Lee, K.H.; Ko, H.C.; Jung, G.Y. Enhancing the charge transfer of the counter electrode in dye-sensitized solar cells using periodically aligned platinum nanocups. *Small* **2012**, *8*, 3757–3761. [CrossRef] [PubMed]

66. Park, K.-H.; Kim, S.J.; Gomes, R.; Bhaumik, A. High performance dye-sensitized solar cell by using porous polyaniline nanotubes as counter electrode. *Chem. Eng. J.* **2015**, *260*, 393–398. [CrossRef]

67. Yue, G.; Wu, J.; Xiao, Y.; Huang, M.; Lin, J.; Lin, J.-Y. High performance platinum-free counter electrode of molybdenum sulfide–carbon used in dye-sensitized solar cells. *J. Mater. Chem. A* **2013**, *1*, 1495–1501. [CrossRef]

68. Zheng, X.J.; Guo, J.H.; Shi, Y.T.; Xiong, F.Q.; Zhang, W.H.; Ma, T.L.; Li, C. Low-cost and high-performance $CoMoS_4$ and $NiMoS_4$ counter electrodes for dye-sensitized solar cells. *Chem. Commun.* **2013**, *49*, 9645–9647. [CrossRef] [PubMed]

Article

Mechanical and Microstructural Features of Plasma Cut Edges in a 15 mm Thick S460M Steel Plate

Javier Aldazabal [1,*], Antonio Martín-Meizoso [1], Andrzej Klimpel [2], Adam Bannister [3] and Sergio Cicero [4]

1   CEIT and Tecnun, University of Navarra, Manuel de Lardizábal 15, 20018 San Sebastián, Spain; ameizoso@ceit.es
2   Politechnika Slaska—Sutil, ul. Komarskiego 18a, 44-100 Gliwize, Poland; andrzej.klimpel@polsl.pl
3   Tata Steel, Swinden Technology Centre, Moorgate, Rotherham S60 3AR, UK; adam.bannister@tatasteel.com
4   LADICIM (Laboratory of Materials Science and Engineering), E.T.S. de Ingenieros de Caminos, Canales y Puertos, University of Cantabria, Av/Los Castros 44, 39005 Santander, Spain; sergio.cicero@unican.es
*   Correspondence: jaldazabal@tecnun.es; Tel.: +34-943-219-877

Received: 4 May 2018; Accepted: 8 June 2018; Published: 11 June 2018

Abstract: In general, the thermal cutting processes of steel plates are considered to have an influence on microstructures and residual stress distribution, which determines the mechanical properties and performance of cut edges. They also affect the quality of the surface cut edges, which further complicates the problem, because in most cases the surface is subjected to the largest stresses. This paper studies the influence of plasma cutting processes on the mechanical behavior of the cut edges of steel and presents the characterization results of straight plasma arc cut edges of steel plate grade S460M, 15 mm thick. The cutting conditions used are the standard ones for industrial plasma cutting. The metallography of CHAZ (Cut Heat Affected Zones) and hardness profiles versus distance from plasma cut edge surface are tested; the mechanical behavior of different CHAZ layers under the cut edge surface were obtained by testing of instrumented mini-tensile 300 μm thick specimens. Also, the residual stress distribution in the CHAZ was measured by X-ray diffraction. The results for the mechanical properties, microstructure, hardness, and residual stresses are finally compared and discussed. This work concludes that the CHAZ resulting from the plasma cutting process is narrow (about 700 μm) and homogeneous in plate thickness.

Keywords: plasma cutting; cut heat affected zone; mini-tensile test; steel plate; residual stress

---

1. Introduction

The use of steel plates in construction elements, structures, parts of machinery, etc. requires, in practically all cases, the cutting of these metallic sheets into smaller parts. These parts will later be connected to other elements using mechanical joints or welds. Nowadays we have at our disposal a plethora of cutting techniques: shear, oxy-cut, laser, plasma, water jet (with or without abrasive particles), thermal lance, etc., but all these cutting techniques introduce modifications in the regions close to the cut surface: they modify their surface roughness and, when they provide enough heat, they introduce modifications in the microstructure [1]. These changes result in local variations in the mechanical properties; in many cases, they also introduce or modify the profiles of the residual stresses in the areas close to the cutting surface [2–5]. A question arises here as to whether it is preferable to leave the cutting edge as it is or, on the contrary, if it is preferable to eliminate it, for example, by grinding (as specified or recommended in some construction standards [6,7]) to optimize the use of the cut pieces and, in particular, their subsequent performance in applications under alternating loads (fatigue) [8–11].

Criteria to establish whether it is better to keep the original edge or to remove it was one of the objectives proposed within the European project HIPERCUT ("High Performance Cut Edges in Structural Steel Plates for Demanding Applications", RSFR-CT-2012-00027 [12]). In this project, the results obtained using different cutting techniques were analyzed. Steel plates with thicknesses ranging from 8 mm up to 25 mm were studied, while the grades and mechanical strength of the plates varied between S355M and S890Q [13–16]. The cutting techniques that were analyzed and compared were plasma jet, laser beam, and oxyfuel (oxy-acetylene). This article presents the results obtained with the plasma cutting technique in a 15 mm S460M structural steel plate.

2. Materials and Methods

The characterization of the cutting edge obtained in a 15 mm thick steel plate with nominal yield stress of 460 MPa is summarized. The chemical composition of the steel was as follows (wt. %): 0.12 C, 0.45 Si, 1.49 Mn, 0.012 P, 0.001 S, 0.062 Cr, 0.001 Mo, 0.016 Ni, 0.048 Al, 0.011 Cu, 0.036 Nb, 0.005 N, 0.002 Sn, 0.003 Ti, 0.066 V. This plate was cut with a plasma jet, in the standard industrial conditions for the cutting of this thickness. The cutting parameters for the cutting equipment used (Hypertherm 260) and for the referred thickness were: Plasma arc current 200 A, arc voltage 131 V, cutting speed 2200 mm/min., torch standoff 4.1 mm, O_2 plasma gas flow rate 69 L/min. and shielding gas flow rate 28 L/min. Here, it should be noted that the objective of the research is not to optimize the cutting parameters, but to determine how well-defined industrial parameters affect the cut edge properties and characteristics.

After the cutting process, samples were obtained (from the cut edge area) for metallographic study, hardness measurements and machining of mini-specimens for tensile tests. The metallographic samples were fixed in a conductive acrylic resin (Condufast™, Struers, Cleveland, OH, USA). Then, the surface under observation was polished with SiC papers up to grade 1200 and finally polished with 0.6 μm diamond paste, on velvet, until a specular finish is achieved. The polished samples were etched with 2% Nital for 15 s, rinsed with ethyl alcohol and dried under a hot air stream, before being observed in an optical microscope (Leica MEF-4, Leica, Wetzlar, Germany).

The hardness profiles were made using a LECO hardness tester (model M-400-G2, Leco, Saint Joseph, MI, USA) equipped with a Vickers pyramidal tip. The indentations were carried out with a load of 4.93 N (0.5 kg).

The mini-tensile samples were cut from the plate using a wire electro-discharge machine (WEDM). Four dog bone shaped prisms were cut from the surface of the cut edge, with their longitudinal axes in the direction of the cut. These prisms were in the mid plane of the plate thickness (or the cut edge), as shown in Figure 1.

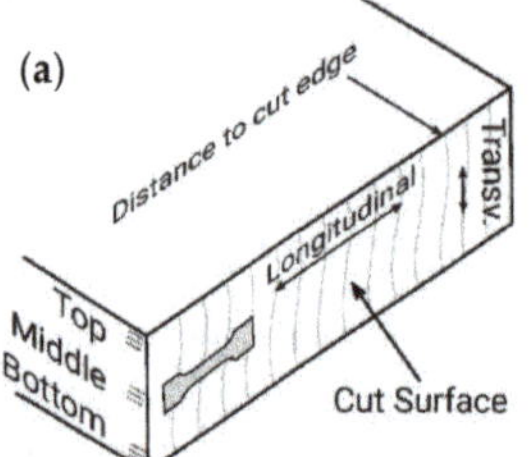

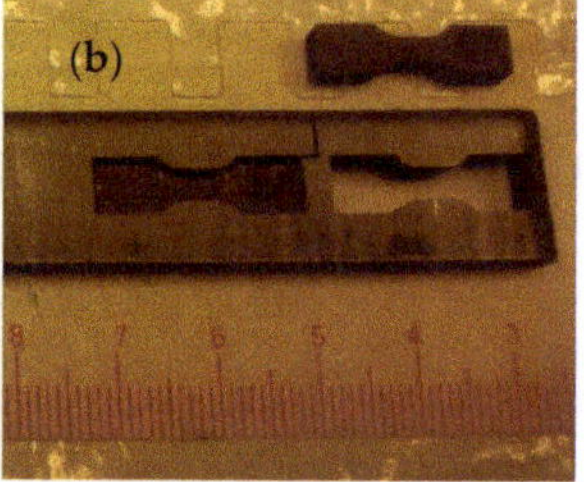

Figure 1. (a) General geometry of the cut plate including the orientation of tensile specimens, zones where hardness measurements were made, and directions used for measuring stresses. (b) Extraction of four bone shaped blocks (for later slicing and extraction of the mini-tensile samples), by electro-erosion of the central zone of the cut edge. Scale shown is in millimeters.

These four prisms were sliced into 300 µm-thick specimens. The distance between two consecutive specimens obtained from the same block was also 300 µm, due to the material removed by the wire during the machining process (see Figure 2). In other words, between two consecutive mini-tensile specimens obtained from the same block, there were 300 µm of material that were lost during the machining process that, consequently, could not be characterized. To better characterize the tensile properties all along the depth from the cut edge, the initial cut of each block or prism was moved (in depth) 150 µm from one another. With this shift between blocks, it was possible to obtain mini-tensile samples each of 150 µm in depth (or distance to the cutting edge).

Figure 2. Slicing of one of the prisms to obtain tensile mini-samples.

As can be seen from Figures 1 and 2, the mini-tensile samples have a longitudinal orientation (same as the cut edge) and their faces are parallel to the cut edge.

According to the bibliography [17], wire EDM cutting (WEDM) introduces residual stresses up to a depth of approximately 80 µm. To eliminate, as much as possible, the effects of the WEDM, 50 µm were eliminated on each side of the mini-tensile samples by polishing (with SiC grade 1200 sandpaper and a subsequent polishing with 1 µm diamond paste in a velvet cloth) [18]. To remove the material, an automatic Struers polishing machine (Struers, Cleveland, OH, USA) was used. During the polishing, the orientation of the samples was fixed to provide a longitudinal polishing pattern, parallel to the future direction of loading. The final thickness of the mini-tensile samples was nominally 200 µm (the real thickness was measured and recorded for each sample. Average sample thickness was 203 µm with a standard deviation of 25 µm).

Exceptionally, in the sample containing the surface of the cutting edge, this surface was not removed by polishing, but it was preserved. This first mini-tensile specimen was polished only on the inner side (polishing was carried out removing 100 µm on the inner side, so that the final thickness was also approximately 177 µm).

Figure 3 shows a mini-tensile sample. Its nominal dimensions are 20 mm in length, 5 mm in total width (the gauge length having 2.5 mm width and 3 mm length) and 0.2 mm in thickness.

To obtain accurate strain measurements, each specimen was instrumented with a strain gauge (HBM 1-LY11-3/120, HBM, Darmstadt, Germany, with 5% maximum strain), as shown in Figure 3. A San-Ei (San-Ei Electric Co., Ltd., Osaka, Japan) amplifier is used to record the lengthening of the strain gauge. The tensile tests were carried out in a universal testing machine with a crosshead displacement speed of 0.1 mm/minute. For deformations greater than 5%, the crosshead position records were used (based on the correlation between the position of the actuator and the previous measurements of the strain gauge). The strain gauge stiffness is not negligible when compared to that of the specimen, and it was measured in an independent test. Thus, the contribution of the strain gauge to the force measured by the load cell was considered when defining the actual load applied to the mini-tensile specimens during tests.

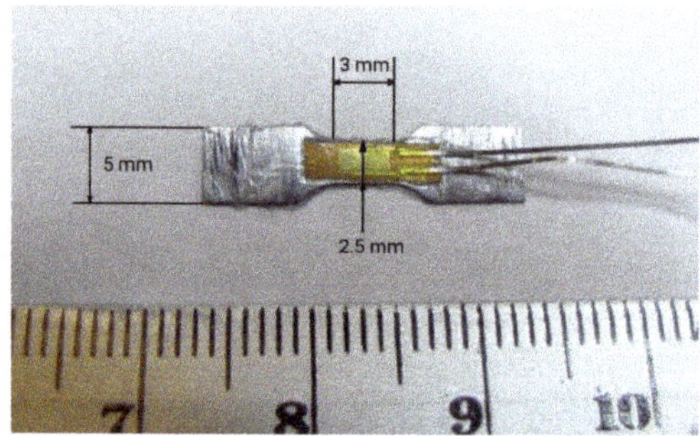

Figure 3. Mini-tensile specimen with the strain gauge for measuring deformations up to 5%. Scale shown is in millimeters.

The tensile tests were carried out in an electro-mechanical machine (Instron Mini 44, Instron, High Wycombe, UK). This test machine is equipped with a load cell of $\pm$500 N. For fixing the samples, their ends were inserted in two narrow channels machined at the end of two bolts. These channels were 300 μm wide. The mini-tensile samples shoulders were fixed into the channels with a cyanoacrylate adhesive (Loctite); the capillarity of glue guarantees the complete fixation of the sample.

Finally, X-ray diffraction equipment was used to measure longitudinal (along the cutting direction) and transversal (thickness direction) residual stresses at different depths from the cut edge. Prismatic samples of $8 \times 10 \times 35$ mm were machined (using WEDM) from half the thickness of the plate to avoid the influence of the top/bottom surfaces. These samples were cleaned in a solution of 500 mm^3 of HCl and 500 mm^3 of distilled water, for 20 min at room temperature. The measurements were made on a X-Ray diffractometer (X'Pert, Philips, Amsterdam, The Netherlands), with the following parameters: anode material Cr (K-α_2 = 2.2936663 Å), voltage 40 kV, current 40 mA, angle 2θ scanning range 144.1~166.0° (0.3°/step), ψ scan range 60.00~60.00° (7.76°/step), time per step 12.05 s. The lattice equivalent planes considered for measuring the residual stresses were the {211}.

For the measurement of stresses at different depths, the cut surface material was eliminated in a controlled manner. To remove these thin layers of steel, an electrolytic polish setup was used. This technique allowed us to remove material without introducing additional stresses in the samples. The material was etched applying a potential of 14 V on blocks in an electrolytic medium made of 90% perchloric acid and 10% ethanol. This procedure (electro-polishing and X-ray diffraction) was repeated four times, until reaching a depth of 700 μm from the original cut surface. The stresses were always measured on the polished surface. The measured residual stress does not correspond to the original stress at that depth because the removal of material induces a relaxation of internal stresses. It is possible to deduce the original stresses at a certain depth, σ, using expression [19]:

$$\sigma(z_1) = \sigma_m(z_1) + 2\int_{z_1}^{H} \frac{\sigma_m(z_1)}{z}dz - 6z_1\int_{z_1}^{H} \frac{\sigma_m(z_1)}{z^2}dz$$

where H corresponds to the initial sample thickness, z_1 to the current thickness and σ_m to the measured stress obtained after eliminating the material. An in-house developed code was used to deduce the stress profile that exists on the original plate from measurements and material thickness removed.

To validate this measurement methodology, a set of 5 tests were carried out on a pre-stressed sample. These 5 samples were stressed up to 390 MPa. Results from X-Ray measurements provided stress values of 388 $\pm$ 7 MPa.

3. Results

3.1. Metallography

Figure 4 shows the metallographic section (already etched) of the zone affected by the cut (CHAZ). In the figure, the left side corresponds to the cut edge, whereas the right side corresponds to the bulk material, or remaining plate. Moreover, the top side of the picture corresponds to the plate surface where the plasma nozzle was located, and the bottom side corresponds to the zone that rested on the cutting table (slag side). Rounding and a thickness reduction of around 1.0 mm are observed in the upper-left zone (jet entry) of the cut. The CHAZ (in principle, the darker material) is very thin, around 0.4 mm deep, especially when compared to that obtained by other cutting procedures with lower energy densities, such as oxyfuel. The latter generates a CHAZ whose depth varies between 1 mm (nozzle side) and 4 mm (slag side) [12,20]. The CHAZ provided by the plasma cut is, however, deeper than that obtained in the same material when cutting with laser, which varies between 0.1 mm (nozzle side) and 0.4 mm (slag side) [12,20]. Moreover, the depth of the CHAZ obtained when performing plasma cuts is nearly constant, whereas oxyfuel and laser cuts generate HAZs with variable depths along the plate thickness.

The heat generated during the cutting process produces phase transformations and the grain growth of the underlying matrix material, as shown in Figure 5. The microstructure shown in this figure corresponds to the area highlighted in the frame in the middle section of Figure 4. Grain size reduces drastically in areas close to the cut edge. At distances from the cut edge greater than approximately 600 µm, there are no changes in gran size.

Just below the cutting edge, layers of martensite and bainite are observed. At a depth of about 200 µm from the cutting edge, polygonal ferrite is observed. At approximately 400–500 µm, the ferrite grains are larger and beyond the 700 µm pearlite and (even larger) polygonal ferrite grains are observed; this last microstructure corresponds to the base material, not affected by the cut (a hypoeutectoid steel, with bands of perlite and ferrite).

3.2. Microhardness

Figure 6 shows the Vickers hardness profiles (0.5 kg, HV05) of the cutting edge measured at the upper part of the plate (at 0.5 mm and 2.5 mm from the plate upper surface, or nozzle side), at half the thickness plate, and at the lower part of the plate (at 0.5 mm and 2.5 mm from the plate lower surface). The measurements are presented as a function of the distance to the cutting edge (see also Figures 1a and 4 to clarify the position of the measurements). To obtain detailed hardness profiles, indentations of 0.5 kg (4.91 N) were made instead of 1 kg (9.81 N). It is known that non-standard hardness measurements can differ from standard ones but the indentations of half a kilogram are smaller and can be placed closer to one to another and to the cutting edge itself. At each height, three lines of indentations were made, with a small shift between them, with the purpose of obtaining in greater detail the evolution of hardness in the CHAZ versus depth. For each depth a single measurement was made.

It can be observed that the microhardness measurements provide very similar results in the upper, middle, and lower part of the cut section. This is related with the uniform thickness of the CHAZ, as revealed in the images included in Figure 4. The effect of plasma cutting vanishes at a distance of approximately 700 µm from the cut edge, along all the cut thickness. The European standard EN 1090-2 [21] sets a limit of 380 kg/mm^2 for the Vickers hardness after cutting. The microhardness measured near the surface slightly exceeds this limit in a thin layer, with a depth of about 350 µm (EN 1090-2 [21] specifies Vickers with 1 kg load and here, the results are presented with only 0.5 kg). The problem with a very hard cut surface is that it is prone to cracking in subsequent bending processes. Several bend tests of the cut edges analyzed in the HIPERCUT project were carried out (e.g., [12,22]). All samples tested were bent 180° without cracks, including those specimens that had the entire surface of the cutting edge on the tensile side, revealing that the observed hardness measurements, which are slightly above the limits provided in [21], do not negatively affect the bending behavior of the plasma cut edges analyzed.

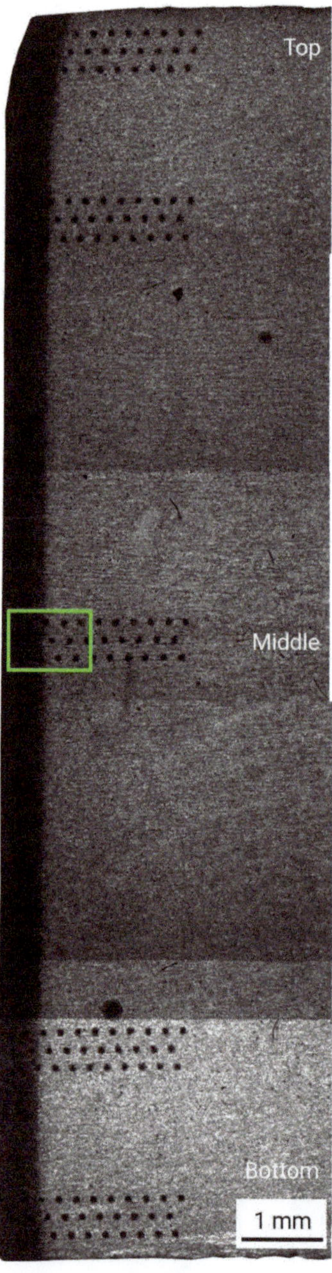

Figure 4. Optical micrograph of the CHAZ, ethed with 2% Nital. The Vickers indentations made for measuring hardness profiles are visible. The upper-left part corresponds to the plasma inlet (see also Figure 1a).

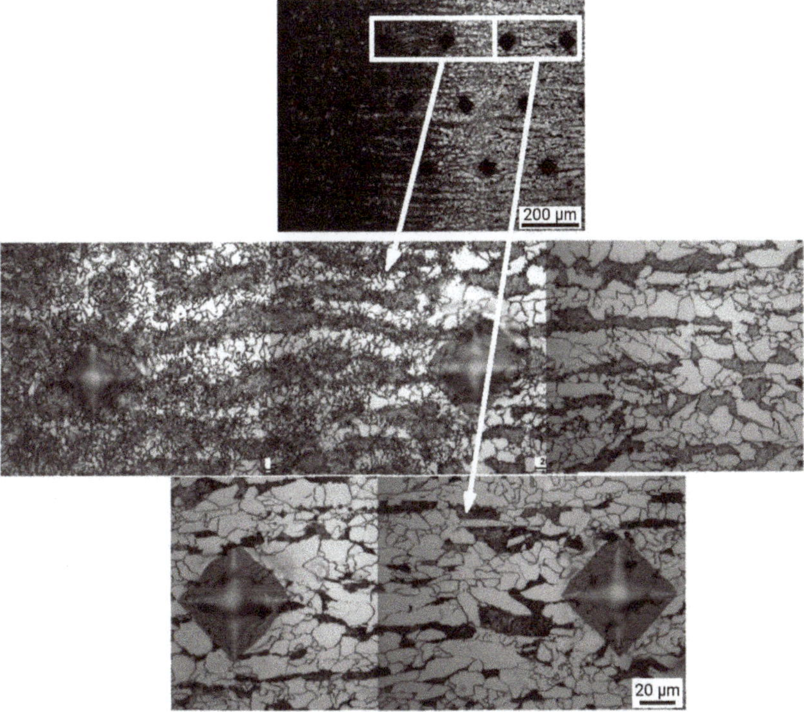

Figure 5. Detail corresponding to the framed area on Figure 4 close to the middle section. The microstructures shown in the second row correspond to the zone closer to the cut edge and its microstructure is finer in regions closer to the cut edge (left).

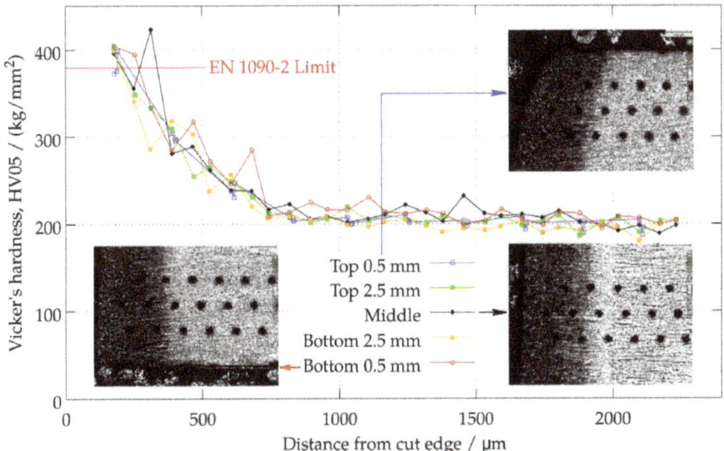

Figure 6. Hardness profiles against the distance to the cut edge. Each color corresponds to a region of the sample, blue corresponding to the one closest to the plasma nozzle and red to the region further from the plasma nozzle. Images included in the graph show the microstructure and indentations distribution used to make the plot in three regions (**top**, **middle** and **bottom**).

3.3. Mini-Tensile Tests

Figure 7 shows the (ductile) fracture observed in a tensile test, typical of a mini-tensile specimen after necking. Figure 8 summarizes the results obtained in the mini-tensile tests. The engineering stress is plotted against the (engineering) strain, as a function of the distance to the original surface produced by the plasma cut. As mentioned above, the mini-tensile specimens were obtained from four blocks extracted from the center of the thickness of the cut edge. For each depth, a single tensile test was carried out. It can be observed that the closer to the cut edge, the larger the resistance parameters and the lower the ductility. Differences in stress-strain curves tend to stabilize at a distance of approximately 750–900 μm. This is basically consistent with the microhardness measurements made at the center of the cutting edge.

Figure 7. Mini-tensile test specimen during tensile test. The top and bottom screws have the channel parallel to the image plane where sample shoulders were glued. The necking and later fracture can be easily observed.

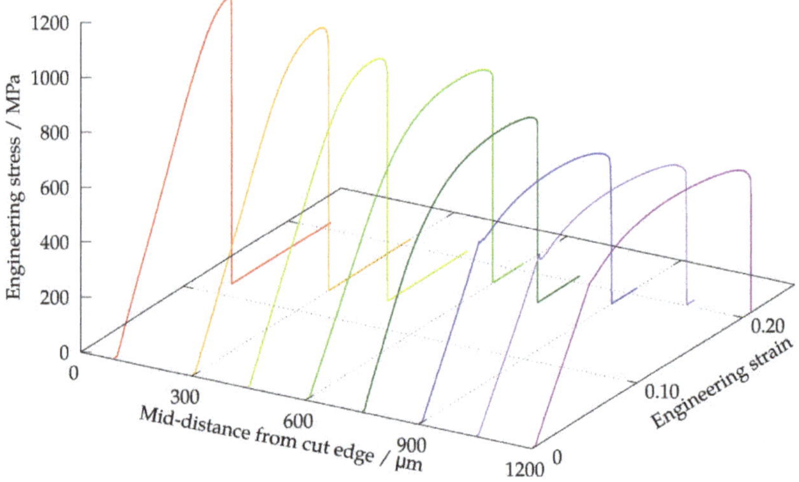

Figure 8. Stress-strain curves obtained at different depths from the cut edge.

Figure 9 represents the change in the mechanical behavior (yield stress, Ultimate tensile strength, etc.) versus distance to plasma cut, for mini-specimens extracted from the central section of the cut. From this figure, it is clear that the material strength decreases as the distance from cut edge increases. For distances of over 800 µm there are no changes in the ultimate tensile strength, UTS, and the mechanical properties of the base material are reached. Something similar happens with the yield stress, which decreases from 1060 MPa at 88.5 µm to 526 MPa at 750 µm. For distances greater than 750 µm there are no changes in UTS, so it could be assumed that base material has been reached.

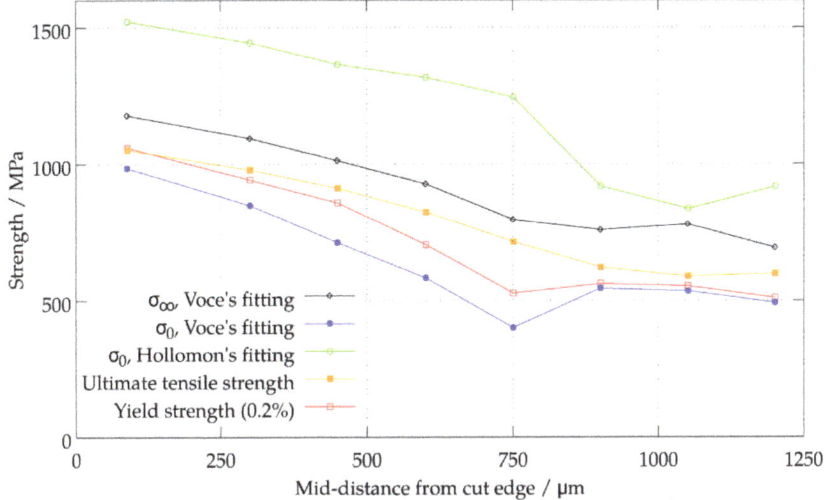

Figure 9. Evolution of mechanical properties as a function of distance to cut edge.

Figure 10 shows the evolution of the uniform strain (up to necking), fracture strain, hardening index, and critical strain for Voce's fitting, versus distance to plasma cut (Appendix A).

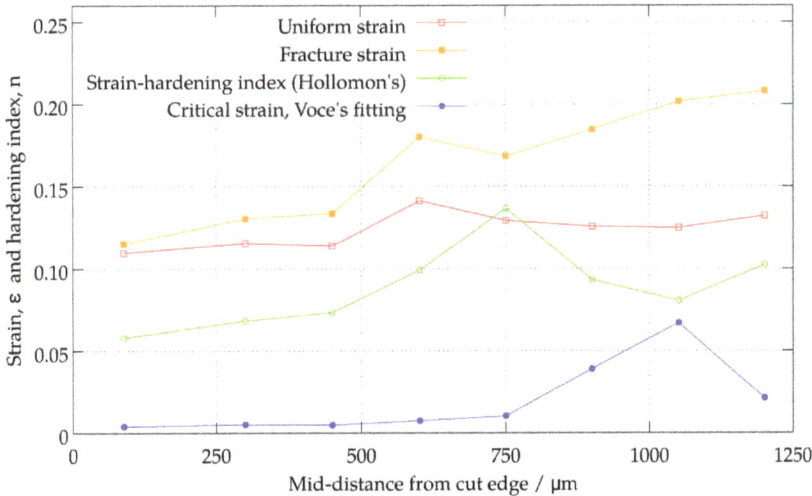

Figure 10. Evolution of the uniform strain, fracture strain, hardening index, and critical strain as a function of the distance to edge cut by plasma.

From this figure, it is possible to appreciate that the material gets softer with the distance from the cut edge. There are no significant changes in the ductility over distance but the fracture strain increases, making the material tougher. From the fitting parameters shown in this figure, it is not possible to say that at 1 mm the base material has been reached, but with the information of Figures 8 and 9 it could be assumed that, from a practical point of view, CHAZ reaches 700 μm. This deduced value is smaller than the one observed for other thermal cutting processes [23].

3.4. Residual Stresses

Figure 11 shows the residual stresses measured in the longitudinal direction (L) and thickness direction (T) obtained by X-ray diffraction, after its deconvolution. Error in depth was obtained by measuring the removed material in different locations of the sample. The error in residual stress was obtained from the error given by the X-Ray equipment used and the error previously measured in depth. Plasma cutting generates at the surface of the cut a great compression in both directions; L and T (see Figure 1a). The residual compression extends to a distance of approximately 700 μm in the underlying material, located under the cut surface.

The values of these compressive residual stresses measured on the surface are in the order of the material yield stress. These big stresses are probably produced by the great cooling gradients that appear in the plate during the cutting process and they can play a key role in the fatigue behavior of these cut edges (for example, delaying the initiation of cracks and, therefore, increasing fatigue life). However, this behavior also depends on other parameters such as the surface roughness and the microstructural characteristics at those areas underlying the cut [14–17]. Whether or not the removal of the plasma cut surface would be beneficial for the material fatigue performance requires further research.

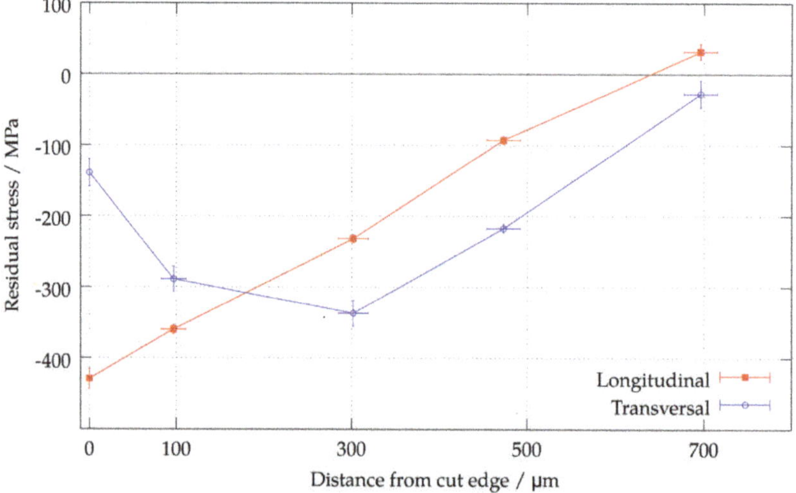

Figure 11. Distribution of residual stresses according to distance to cut edge.

4. Conclusions

- The Cut Heat Affected Zone (CHAZ) generated by plasma cutting is quite narrow (around 700 μm and, in any case, lower than 1 mm) and quite uniform across the entire thickness of the cut. These results agree with metallographic observations and microhardness measurements.
- The hardness measured (HV05) on the surface of the cutting edge is slightly higher than the limit set in the standard EN 1090-2 [21] (although it does not seem to affect bend behavior).

- It was possible to obtain Stress-strain curves for the material at different depths under the cut. This was done by testing mini-tensile specimens, extracted by means of WEDM, and instrumented with strain gauges.
- The yield stress and the UTS change with the distance to the cut edge. In the areas closest to the cut, values are obtained that can be 100% higher than those measured at the base material.
- The greater the mechanical strength, the lower the ductility and resilience. Obviously, their values are related to those microstructures formed in the CHAZ.
- Plasma cutting introduces large residual compressive stresses, up to depths of approximately 700 μm.

Author Contributions: J.A. wrote the paper, performed the measurements of residual stresses and analyzed X-Ray results. A.M.M. designed mini-tensile tests, prepared the samples, and performed the experiments, analyzing their results. He also contributed with the manuscript. S.C. contributed with the writing of the paper and designed, programmed and analyzed all indentation experiments. A.K. supplied all the samples and performed the bending tests. He also optimized conditions for performing plasma the cuts. A.B. performed the metallography preparation and analysis of the cut samples.

Acknowledgments: The authors of this work would like to express their gratitude to the European Union for the financial support of the project HIPERCUT: "High Performance Cut Edges in Structural Steel Plates for Demanding Applications" (RSFR-CT-2012-00027), on the results of which this paper is based.

Conflicts of Interest: The authors declare no conflict of interest.

Appendix A

The simplest fit to tensile behavior (stress vs. strain curve) of a material is given by Hollomon's expression [24]

$$\sigma = \sigma_0 \varepsilon_p^n \tag{A1}$$

where σ represents the stress as a function of the plastic strain, ε_p. Only two parameters are used: σ_0 and n. n is known as strain hardening index.

A little more sophisticated and realistic is Voce's relation [25];

$$\sigma = \sigma_\infty - (\sigma_\infty - \sigma_0)e^{-\frac{\varepsilon_p}{\varepsilon_c}} \tag{A2}$$

This expression has three parameters: σ_∞ represents the saturation stress (the stress that ideally will be reached for infinite strain), σ_0 is the stress for a negligible plastic deformation and σ_c is a critical strain (as shown in Figure A1, at intercept with the abscissa).

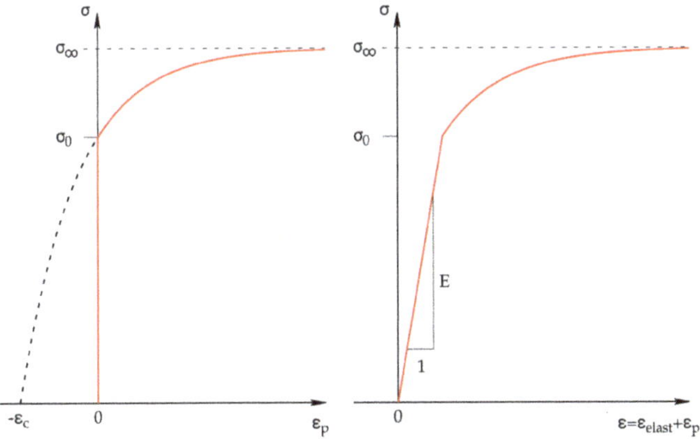

Figure A1. Interpretation of Voce's parameters.

References

1. Wood, W.E. *Heat-Affected Zone Studies of Thermally cut Structural Steels (Report FHWA-RD-93-O 15)*; US Department of Transportation Federal Highway Administration: Washington, DC, USA, 1994.
2. Tomas, D.J. Characterisation of Steel Cut Edges for Improved Fatigue Property Data Estimations and Enhanced CAE Durability. Ph.D. Thesis, Swansea University UK, Swansea, UK, 2011.
3. British Standard EN ISO 9013:2002. *Thermal Cutting—Classification of Thermal Cuts—Geometrical Products Specification and Quality Tolerances*; British Standards Institution: London, UK, 2004.
4. Kirkpatrick, I. Variety of cutting processes spoil fabricators for choice. *Weld. Met. Fabr.* **1994**, *62*, 11–12.
5. Avila, M. Which metal-cutting process is best for your application. *Weld. J.* **2012**, *91*, 32–36.
6. The Steel Construction Institute. *Guidance Notes on Best Practice in Steel Bridge Construction*; P185, Fifth Issue; Steel Bridge Group: Ascot, UK, 2010; ISBN 978-1-85942-196-3.
7. *EN 1993-1-9: Eurocode 3: Design of Steel Structures—Part 1–9: Fatigue*; European Committee for Standardization: Brussels, Belgium, 2005.
8. Goldber, F. Influence of thermal cutting and its quality on the fatigue strength of steel. *Weld. J.* **1973**, *52*, 392–404.
9. Plecki, R.; Yeske, R.; Alstetter, C.; Lawrence, F.V., Jr. Fatigue resistance of oxygen cut steel. *Weld. J.* **1977**, *56*, 225–230.
10. Ho, N.-J.; Lawrence, F.V.; Alstetter, C.J. The fatigue resistance of plasma- and oxygen-cut steel. *Weld. Res.* **1981**, *11*, 231–236.
11. Piraprez, E. *Fatigue Strength of Flame-Cut Plates, Fatigue of Steel and Concrete Structures*; IABSE Reports; International Association for Bridge and Structural Engineering: Zurich, Switzerland, 1982; Volume 37, pp. 23–26.
12. Bannister, A.; Danks, S.; Klimpel, A.; Luksa, K.; Rzeznikiewicz, A.; Cicero, S.; Álvarez, J.A.; García, T.; Martín-Meizoso, A.; Aldazabal, J. *High Performance Cut Edges in Structural Steel Plates for Demanding Applications (HIPERCUT)*; EUR 28092 EN; Directorate-General for Research and Innovation, European Commission: Luxembourg, 2016; ISBN 978-92-79-61684-6.
13. Kaufmann, I.; Schonherr, W.; Sonsino, C.M. Fatigue strength of high-strength fine-grained structural steel in the flame cut condition. *Schweissen u Scheniden* **1995**, *3*, E46–E51.
14. Cicero, S.; García, T.; Álvarez, J.A.; Martín-Meizoso, A.; Aldazabal, J.; Bannister, A.; Klimpel, A. Definition and validation of Eurocode 3 FAT classes for structural steels containing oxy-fuel, plasma and laser cut holes. *Int. J. Fatigue* **2016**, *87*, 50–58. [CrossRef]
15. Cicero, S.; García, T.; Álvarez, J.A.; Bannister, A.; Klimpel, A.; Martín-Meizoso, A.; Aldazabal, J. Fatigue behaviour of structural steels with oxy-fuel, plasma and laser cut straight edges. Definition of Eurocode 3 FAT classes. *Eng. Struct.* **2016**, *111*, 152–161. [CrossRef]
16. Cicero, S.; García, T.; Álvarez, J.A.; Martín-Meizoso, A.; Bannister, A.; Klimpel, A. Definition of BS7608 fatigue classes for structural steels with thermally cut edges. *J. Constr. Steel Res.* **2016**, *120*, 221–231. [CrossRef]
17. García Navas, V.; Ferreres, I.; Marañon, J.A.; Garcia-Rosales, C.; Gil Sevillano, J. Electro-discharge machining (EDM) versus hard turning and grinding—Comparison of residual stresses and surface integrity generated in AISI O1 tool steel. *J. Mater. Process. Technol.* **2008**, *195*, 186–194. [CrossRef]
18. Rech, J.; Kermouche, G.; Grzesik, W.; García-Rosales, C.; Khellouki, A.; García-Navas, V. Characterization and modelling of the residual stresses induced by belt finishing on a AISI52100 hardened steel. *J. Mater. Process. Technol.* **2008**, 187–195. [CrossRef]
19. Moore, G.; Evans, W.P. Mathematical Correction for Stress in Removed Layers in X-ray Diffraction Residual Stress Analysis. *SAE Trans.* **1958**, 340–345. [CrossRef]
20. Andrés, D.; García, T.; Cicero, S.; Lacalle, R.; Álvarez, J.A.; Martín-Meizoso, A.; Aldazabal, J.; Bannister, A.; Klimpel, A. Characterization of heat affected zones produced by thermal cutting processes by means of Small Punch tests. *Mater. Charact.* **2016**, *119*, 55–64. [CrossRef]
21. European Standard: EN 1090-2:2011+A1. *Execution of Steel Structures and Aluminium Structures. Part 2: Technical Requirements for Steel Structures*; European Committee for Standardization: Brussels, Belgium, 2011.

22. Klimpel, A.; Cholewa, W.; Bannister, A.; Luksa, K.; Przystalka, P.; Rogala, T.; Skupnik, D.; Cicero, S.; Martín-Meizoso, A. Experimental investigations of the influence of laser beam and plasma arc cutting parameters on edge quality of high-strength low-alloy (HSLA) strips and plates. *Int. J. Adv. Manuf. Technol.* **2017**, *92*, 699–713. [CrossRef]
23. Martín-Meizoso, A.; Aldazabal, J.; Pedrejón, J.L.; Moreno, S. Resilience and ductility of Oxy-fuel HAZ cut. *Frattura ed Integrità Strutturale* **2014**, *30*, 14–22. [CrossRef]
24. Hollomon, J.H. Tensile deformation. *Trans. AIME* **1945**, *162*, 268–277.
25. Voce, E. The Relationship between Stress and Strain for Homogeneous Deformation. *J. Inst. Met.* **1948**, *74*, 537–562.

Article

Crystal Structures of GaN Nanodots by Nitrogen Plasma Treatment on Ga Metal Droplets

Yang-Zhe Su and Ing-Song Yu *

Department of Materials Science and Engineering, National Dong Hwa University, Hualien 97401, Taiwan;
s1992811227@yahoo.com.tw
* Correspondence: isyu@gms.ndhu.edu.tw; Tel.: +886-3-890-3219

Received: 20 April 2018; Accepted: 30 May 2018; Published: 4 June 2018

Abstract: Gallium nitride (GaN) is one of important functional materials for optoelectronics and electronics. GaN exists both in equilibrium wurtzite and metastable zinc-blende structural phases. The zinc-blende GaN has superior electronic and optical properties over wurtzite one. In this report, GaN nanodots can be fabricated by Ga metal droplets in ultra-high vacuum and then nitridation by nitrogen plasma. The size, shape, density, and crystal structure of GaN nanodots can be characterized by transmission electron microscopy. The growth parameters, such as pre-nitridation treatment on Si surface, substrate temperature, and plasma nitridation time, affect the crystal structure of GaN nanodots. Higher thermal energy could provide the driving force for the phase transformation of GaN nanodots from zinc-blende to wurtzite structures. Metastable zinc-blende GaN nanodots can be synthesized by the surface modification of Si (111) by nitrogen plasma, i.e., the pre-nitridation treatment is done at a lower growth temperature. This is because the pre-nitridation process can provide a nitrogen-terminal surface for the following Ga droplet formation and a nitrogen-rich condition for the formation of GaN nanodots during droplet epitaxy. The pre-nitridation of Si substrates, the formation of a thin SiN_x layer, could inhibit the phase transformation of GaN nanodots from zinc-blende to wurtzite phases. The pre-nitridation treatment also affects the dot size, density, and surface roughness of samples.

Keywords: nitrogen plasma; Ga droplet; GaN nanodot; transmission electron microscopy; wurtzite; Zinc-blende

1. Introduction

Group-III nitride-based semiconductors (InN, GaN, and AlN) have very wide bandgaps from 0.64 to 6.2 eV for various applications in optoelectronics and electronics [1]. Gallium nitride, with a direct bandgap of 3.4 eV, which shows high carrier mobility and high thermal conductivity, has been successful applied in commercialized devices, such as light emitting diodes (LED) and high electron mobility transistors (HEMT) [2,3]. Moreover, for the quantum computing applications, nanostructures of GaN have large exciton binding energy and confinement potential for devices, such as single-photon emitters [4] and single-electron transistors [5]. In particular, GaN quantum dots are promising materials in these devices [6].

For the fabrication techniques of GaN, GaN thin films can be epitaxially grown by molecular beam epitaxy (MBE), metal-organic chemical vapor deposition (MOCVD), pulsed laser deposition (PLD), and Hydride vapor-phase epitaxy (HVPE) [7–11]. Once the epi-layers have the lattice mismatch with two-dimensional wetting layers or substrates, self-assembled GaN nanodots can form due to the strain relaxation; this method is called the Stranski-Krastanov (SK) mode of epitaxial growth [12,13]. Among these techniques, MBE procedure is performed in an ultra-high vacuum chamber in order to minimize contamination and in a lower-temperature growth condition. Other advantages of MBE are its capability to create heterostructures with sharp interfaces, and also to form metastable phase as zinc-blende structure of GaN. For another growth method of semiconductor nanostructures, droplet epitaxy mode using an MBE system was first proposed by Koguchi in 1990 [14,15]. Therefore, the study of droplet epitaxy technique for GaN nanodots was initiated, which is the method of first forming Ga metal droplets in ultra-high vacuum, followed by the treatment of a nitrogen plasma source [16]. There are some advantages of droplet epitaxy by plasma-assisted MBE systems. For instance, self-organized GaN nanodots can be grown directly on various substrates. Density of GaN nanodots can be controlled by the growth parameters. Metastable zinc-blende structure can be performed in the GaN nanodots on sapphires by droplet epitaxy [17–20].

For the crystal structure of GaN nanostructures, wurtzite GaN is the thermodynamically stable phase, which suffers from the presence of a large built-in electric field, called the quantum confined Stark effect, that may degrade the device performance [21,22]. On the other hand, the metastable zinc-blende (cubic) phase of GaN has no polarization fields. The polarization field effect, called the piezoelectric field effect, significantly affects the band structures and optical gain of optoelectronic devices. The radiative recombination time of cubic GaN quantum dots is two orders of magnitude higher than the one for wurtzite GaN quantum dots. The mobility of electrons and holes in a zinc-blende GaN is also intrinsically higher than in a wurtzite GaN due to lower phonon scattering in cubic crystals [23]. Up to now, most of the reports on the growth of cubic GaN nanodots have been proposed by the method of SK mode [24–26]. For the growth mode of droplet epitaxy, Wang et al. reported the mixture phases of wurtzite and zinc-blende on a sapphire substrate at the substrate temperature of 710 °C [18]. As's group reported cubic GaN quantum dots grown on 3C-AlN (001) substrates [27]. Studies on the phase transformation of GaN nanodots on silicon by droplet epitaxy technique are still rare. In this work, we focus on the investigation the crystal structures of GaN nanodots on Si (111) substrates fabricated by nitrogen plasma nitridation on the Ga metal droplets. The microstructure and crystal structures of GaN nanodots were characterized by transmission electron microscopy (TEM). The growth parameters of GaN nanodots, such as substrate temperatures and surface pre-nitridation treatment, influence on the crystal phases of GaN nanodots. Nonpolar GaN nanodots with cubic crystal structure can be performed on Si substrates by droplet epitaxy for the future applications in solid state quantum devices.

2. Materials and Methods

GaN nanodots grown by the method of droplet epitaxy were carried out in our ULVAC MBE system with a radio frequency (RF) nitrogen plasma source [19]. Two inch Si (111) wafers were cleaned using acetone to remove organic impurities and using 10% HF solution to remove the native oxide. After the chemical cleaning, one Si wafer was immediately put into MBE chamber with the base pressure of 1.0×10^{-7} Pa. Then, Si substrates were heated to the temperature 850 °C for 20 min. After the thermal cleaning, the temperature of Si substrate was set to 500 or 550 °C for droplet epitaxy. For the droplet epitaxy process, nanoscale Ga metal droplets were initially formed on Si substrate at beam equivalent pressure 1.9×10^{-4} Pa for the duration of 1 min by a Knudsen cell with 99.999999% Ga metal. Ga adatoms had high sticking coefficients on the substrate at lower growth temperature. At elevated temperatures, surface Ga accumulation was less effective since Ga atoms can evaporate from the surface [28]. The accumulation of Ga atoms had high surface free energy to form nanoscale Ga droplets according to the Volmer-Weber (VW) growth model [29]. Subsequently, the droplets were converted into GaN nanodots by using a nitrogen plasma source, called nitridation process. The parameters of nitridation were operated at an RF forward power of 500 W and N_2 flux of 2 sccm for 5 or 10 min. In addition, the pre-nitridation treatment at temperature 600 °C for 60 min on Si wafers was optionally conducted before the thermal evaporation of Ga metal droplets. The growth parameters of four samples are summarized in Table 1. MBE system was equipped with an in-situ reflection high-energy electron diffraction (RHEED), which was employed for the observation of surface condition during the process of droplet epitaxy. RHEED patterns of samples C1 and C2 are shown in Figure 1. The flat and reconstructed Si (111) surfaces can be obtained after thermal cleaning in Figure 1a,d, which showed long-streak patterns. After the pre-nitridation treatment on Si, the RHEED patterns changed from striped to foggy due to the formation of amorphous nitride layer on Si, shown in Figure 1b. After the formation of Ga droplets, the pattern of surface was still foggy (not shown) which indicated an amorphous structure on the surface. Then, nitridation process led to the formation of GaN nanodots which showed the ring-centered pattern in Figure 1c. For the sample without the pre-nitridation process (C2), the RHEED pattern became foggy after the formation of Ga droplets, shown in Figure 1e. After the nitridation process, the formation of GaN nanodots also showed the ring-centered pattern in Figure 1f, indicating the polycrystalline GaN nanodots on Si [30].

Table 1. The growth parameters of four samples.

Sample	C1	C2	C3	C4
Pre-nitridation	Yes	No	Yes	No
Substrate temperature	500 °C	500 °C	550 °C	550 °C
Nitridation time	5 min	5 min	10 min	10 min

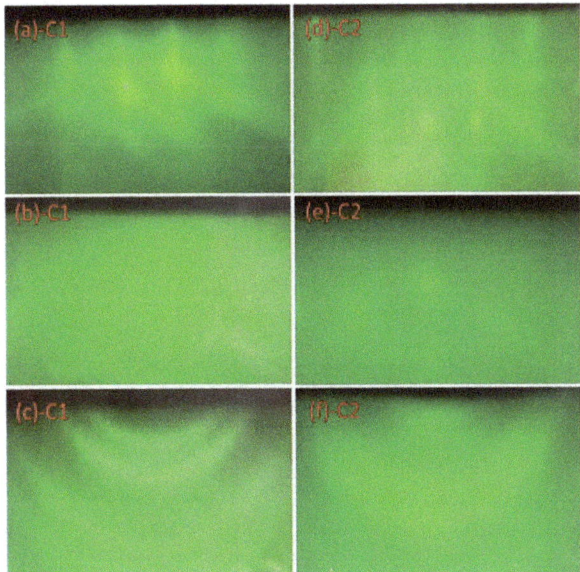

Figure 1. In-situ reflection high-energy electron diffraction (RHEED) observations: (**a**) C1 sample after thermal cleaning; (**b**) C1 sample after pre-nitridation on Si surface; (**c**) C1 sample after nitridation process (i.e., the formation of GaN nanodots); (**d**) C2 sample after thermal cleaning; (**e**) C2 sample after Ga droplets; and (**f**) C2 sample after nitridation process (i.e., the formation of GaN nanodots).

After the formation of GaN nanodots, surface morphology of the samples was examined using the images generated by field-emission scanning electron microscopy (SEM, JSE-7000F, JEOL, Tokyo, Japan) with accelerating voltage 15 KV. According to the observation of SEM, densities of GaN nanodots could be obtained. The surface roughness of samples was investigated by atomic force microscopy (AFM, C3000, Nanosurf, Liestal, Switzerland). The surface chemical composition of samples was studied by X-ray photoelectron spectroscopy (XPS, K-Alpha, Thermo Scientific, Waltham, MA, USA). For the observation of GaN nanodots microstructure and crystal structures, high-resolution TEM images, and selective-area diffraction patterns were performed by using JEOL JEM-3010 with accelerating voltage 300 KV. The TEM specimen were prepared by Ar ion milling with an accelerating voltage of 5 KeV for the cross-section and plane-view TEM observations.

3. Results and Discussion

In the left column of Figure 2, SEM images showed GaN nanodots on the surface of Si (111) in the magnification of 50,000, and their densities in the area of 5.65 μm^2 were calculated from the SEM images as 5.72×10^{10} cm^{-2}, 2.57×10^{10} cm^{-2}, 5.61×10^{10} cm^{-2}, and 5.01×10^{10} cm^{-2} for samples C1, C2, C3, and C4. To compare C1 with C2 or C3 with C4, the pre-nitridation treatment of Si surface made the density of GaN nanodots increase. C1 and C3 samples with pre-nitridation had higher density than the C2 and C4 samples without pre-nitridation, respectively. Moreover, the cross-section TEM images of GaN nanodots in the range of 500 nm are shown in the right column of Figure 2. The average diameters of GaN nanodots were calculated from the TEM images as 21.9 nm, 24.5 nm, 15.1 nm, and 18.3 nm for C1, C2, C3, and C4, respectively. The pre-nitridation treatment made the size of GaN nanodots decrease slightly. In brief, the surface conditions of Si (111) can be modified by the substrate pre-nitridaiton process, as shown in the observation of in-situ RHEED, the nitrogen-terminal surface could influence the sticking coefficient of Ga atoms or the surface diffusion of Ga atoms. The surface-sticking coefficients of Ga atoms over N atoms can be higher than Ga atoms over Si atoms according to the chemical surface stability. The stronger bonding could reduce the surface diffusion length of Ga atoms during the formation of Ga droplets [31,32]. Therefore, we found an increase of density and a decrease of average diameter of GaN nanodots for the samples with nitrogen plasma pre-nitridation treatment (C1 and C3).

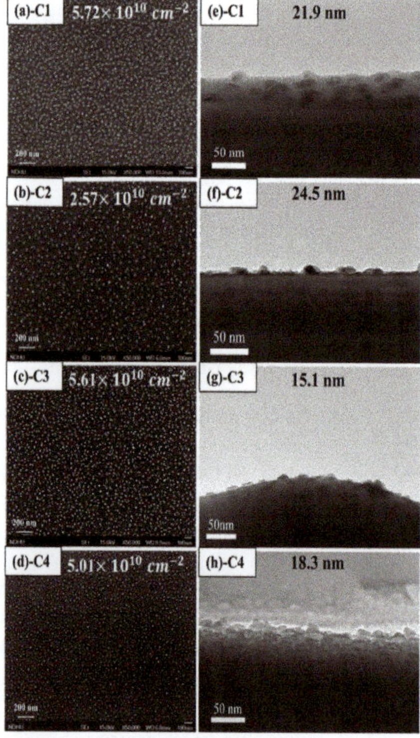

Figure 2. SEM images (**left column**) and cross-section TEM images (**right column**) of GaN nanodots for samples C1, C2, C3 and C4.

To further analyze their surface roughness, AFM images in an area of 5 μm × 5 μm of four samples are shown in Figure 3. The average roughness of samples C1, C2, C3, and C4 are 2.56, 3.04, 1.34, and 1.65 nm, respectively. The samples without pre-nitridation treatment (C2 or C4) had higher surface roughness value than the ones with pre-nitridation (C1 or C3). This is because the larger size and lower density of GaN nanodots on Si which can perform higher surface roughness of samples. The observation of AFM was consistent to the results of SEM and TEM in Figure 2.

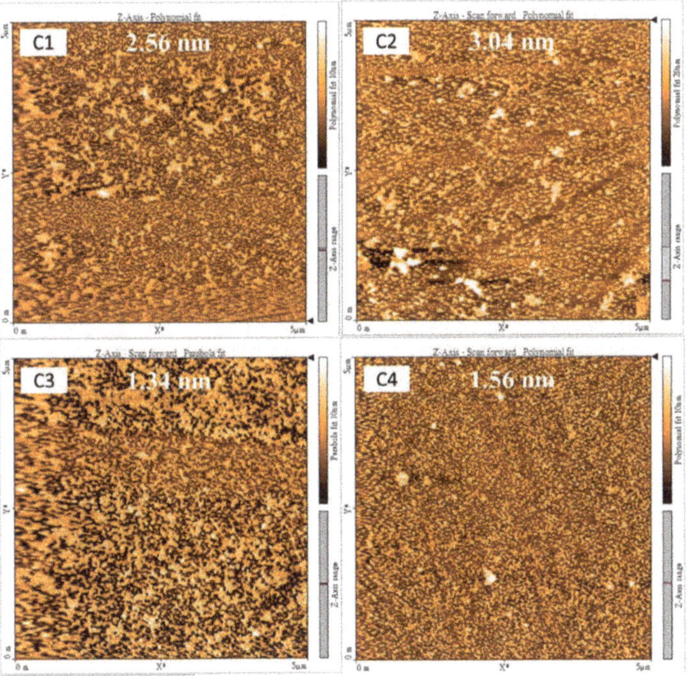

Figure 3. Atomic force microscopy (AFM) images and average roughness of four samples: C1, C2, C3, and C4.

Since the pre-nitirdation treatment of Si (111) influenced the growth of GaN nanodots, we went forward to investigate the surface chemical composition using the measurements of XPS. Figure 4 shows the de-convoluted Ga-3d XPS spectra of GaN nanodots on Si (111) for samples C1, C2, C3, and C4. The XPS spectra were divided into three major components: Ga–O bonding with peak energy around 20.7 eV, Ga–N bonding with peak energy around 19.7 eV, and O–O bonding with the peak energy around 25.0 eV, obtained using the software *Thermo Avantage* (version 4, Thermo Scientific, Waltham, MA, USA). The samples with pre-nitridation treatment (C1 and C3) had relatively stronger peak intensities of Ga–N bonding due to more nitrogen provided for the formation of GaN nanodots. Meanwhile, we found high amount of oxygen chemisorptions on the surface of samples [33]. The strong Ga–O and weak O–O bonds came from the oxidation of nanodots on exposure to the air due to strong difference the electro negativity between Ga and O atoms [21]. This was because the uncompleted crystallization of Ga metal droplets could have existed after nitrogen plasma nitridation. When the nitridation time increased for samples C3 and C4, the peak intensity of Ga–N bonding was stronger due to the formation of the GaN crystal being more complete.

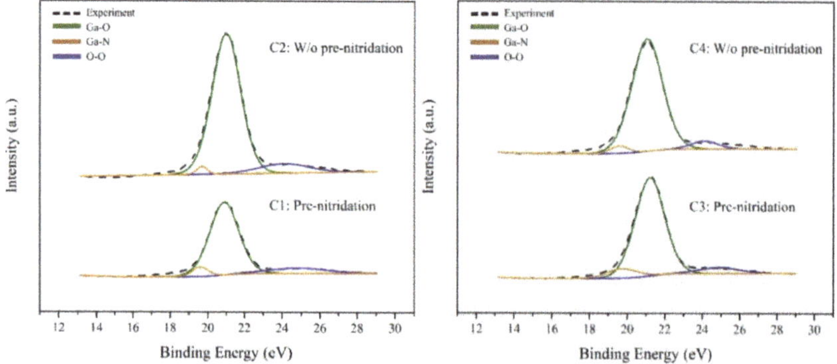

Figure 4. De-convoluted Ga-3d X-ray photoelectron spectroscopy (XPS) spectra of GaN nanodots on Si (111) for samples C1, C2, C3, and C4.

In order to identify the crystal structures of GaN nanodots, we prepared the specimens for the cross-section images of high-resolution transmission electron microscopy (HRTEM) and analyzed the crystal planes using the software *Digital Micrograpy* (version 1, Gatan, Pleasanton, CA, USA). The fast Fourier transformation (FFT) of the original HRTEM images was conducted on a single GaN nanodot, which can provide the simulated diffraction patterns. And then, the inverse fast Fourier transformation (IFFT) of the simulated diffraction patterns provided a clearer crystal structure of a single GaN nanodot. Finally, crystal planes can be identified according to the database of software [34]. Moreover, for the TEM observations of multiple GaN nanodots, we prepared the specimens for plane-view TEM images as shown in Figure 5 and analyzed the selective area diffraction (SAD) patterns of GaN nanodots at a magnification of 500,000. The d-space of ring-like patterns can provide the planes for each ring using the software *CSpot* (version 1, CrystOrient, Zabierzów, Poland) with the aid of a crystallography open database [35]. Figure 5 shows the plane-view TEM image of sample C1 and its ring-like diffraction pattern in the inset. The ring-like pattern came from the polycrystalline GaN nanodots, which was consistent with the results of RHEED observations. In the TEM images, we could also find that the individual GaN dot could contain different misoriented nano-crystals.

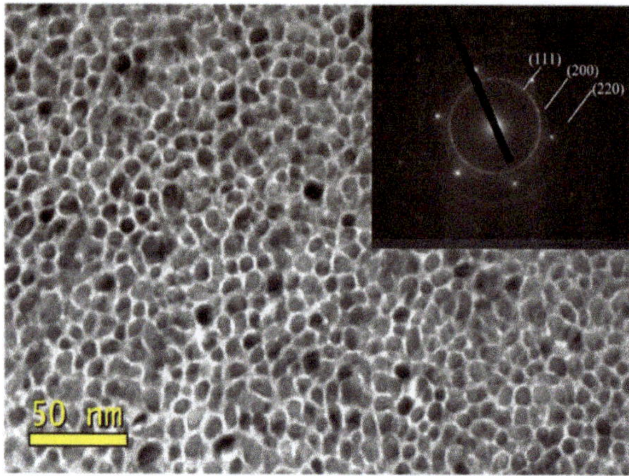

Figure 5. Plane-view TEM images of sample C1 and its diffraction pattern in the inset.

Figure 6a,b shows cross-section HRTEM images of a single GaN nanodot, and Figure 6c,d shows the SAD patterns of GaN nanodots for samples C1 and C2, respectively. According to the analysis using TEM software, the crystal planes of a single GaN nanodot could be identified, and the phase of GaN nanodots could be investigated by the polycrystalline ring-like patterns. For the sample C1, the (200) plane of zinc-blende GaN is shown in Figure 6a, and the cubic structure of GaN nanodots with planes (111), (200), and (220) is shown in Figure 6c. Cubic GaN nanodots could be obtained on Si (111) by the growth parameters of sample C1. On the contrary, we verified the TEM images of sample C2. The (101) and (002) planes of wurtzite structure are shown in Figure 6b, and the ring patterns with planes (100), (002) and (101) for the wurtzite crystal structure of GaN nanodots are shown in Figure 6d. According to the TEM results of samples C1 and C2, the pre-nitridation treatment on Si (111) could inhibit the phase transformation from metastable zinc-blende to stable wurtzite structures. For the microstructures of GaN films grown by MBE, the nitrogen-rich growth condition showed the existence of zinc-blende GaN phase [36]. The pre-nitridation treatment could serve as a larger nitrogen source for the formation of GaN nanodots. Therefore, the cubic GaN nanodots dominated in the sample C1.

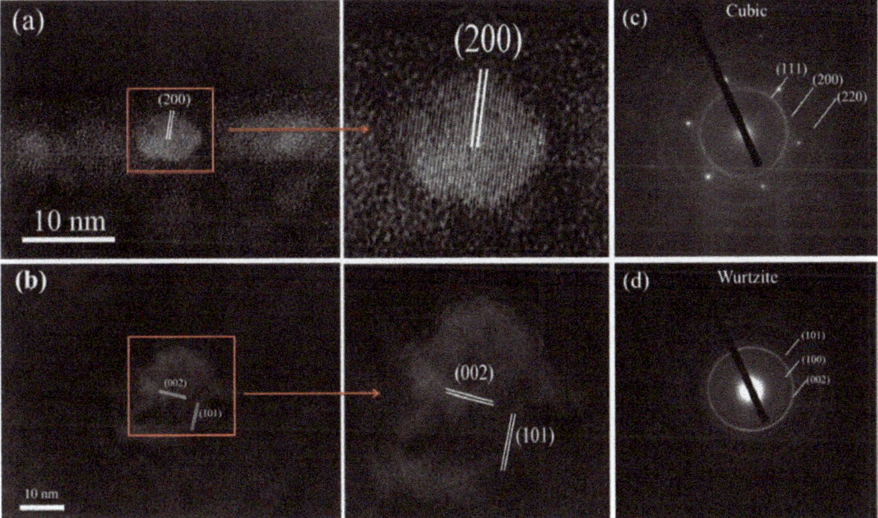

Figure 6. Cross-section HRTEM images of a single GaN nanodot: Sample with pre-nitridation C1 (**a**) and sample without pre-nitridation C2 (**b**); Selective-area diffraction patterns of GaN nanodots: sample C1 (**c**), and sample C2 (**d**).

To further investigate the effect of thermal budget on the crystal structures of GaN nanodots, Figure 7a,b shows cross-section HRTEM images of a single GaN nanodot, and Figure 7c,d shows the SAD patterns of GaN nanodots for samples C3 and C4, respectively. The (002) and (102) planes of wurtzite structure are shown in Figure 7a,b, and the ring-like patterns with planes (100), (002) and (101) both show the wurtzite crystal structure of GaN nanodots in Figure 7c,d. The higher substrate temperature of droplet epitaxy could lead to GaN nanodots phase transformation to wurtzite crystal structure. When the substrate temperature was raised during droplet epitaxy, high surface migration occurred, which improved the quality of the growth front of GaN and it became favorable to form the more thermodynamically equilibrated wurtzite phase [31,36].

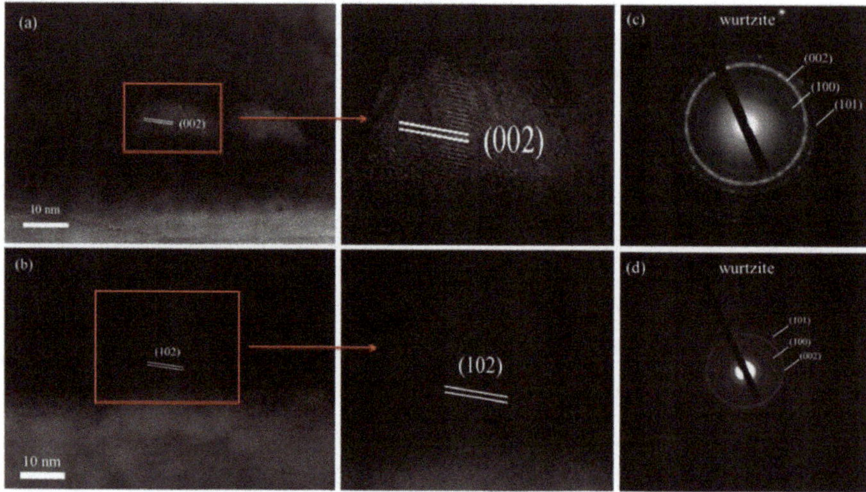

Figure 7. Cross-section HRTEM images of a single GaN nanodot: sample C3 (**a**) and sample C4 (**b**); Selective-area diffraction patterns of GaN nanodots: sample C3 (**c**) and sample C4 (**d**).

4. Conclusions

GaN nanodots were fabricated on Si (111) by droplet epitaxy using a nitrogen plasma-assisted MBE system. Polycrystalline GaN nanodots were shown by the characterizations of in-situ RDEED and TEM. Uncompleted crystallization of the Ga metal droplets were observed from the measurement using XPS. The phase transformation of GaN nanodots from metastable zinc-blende crystal structure to stable wurtzite crystal structure could be inhibited by the surface modification of nitrogen plasma (i.e., pre-nitridation treatment). The pre-nitridation treatment also influenced the size and density of GaN nanodots. The nitrogen-terminal Si surface could not only change the sticking coefficient or the surface diffusion of Ga atoms, but also could provide the nitrogen-rich condition for the existence of zinc-blende phase. On the other hand, the stable wurtzite structure of GaN nandots could be found at higher growth temperatures. This was because the phase transformation of crystal structures depended on the activation free energy barrier. The higher temperature growth could decrease the energy barrier of phase transformation. In this report, we focused on the investigation of GaN-nanodot crystal structures. Nonpolar GaN nanodots with cubic crystal structure could be synthesized during the process of droplet epitaxy, which could have future applications of GaN nanodots for solid state quantum devices.

Author Contributions: I.-S.Y. designed the experiments and wrote the manuscript; Y.-Z.S. performed the experiments and analyzed the data.

Funding: This research was funded by Ministry of Science and Technology Taiwan (MOST 106-2221-E-259-010) and ULVAC Taiwan (NDHU 106A605).

Acknowledgments: The authors acknowledge Ministry of Science and Technology Taiwan and ULVAC Taiwan for financially supporting this study and publication. The authors also would like to acknowledge Micheal Chen and Stanley Wu of ULVAC Taiwan for the maintenance of PA-MBE system.

Conflicts of Interest: The authors declare no conflicts of interest.

References

1. Wu, J. When group-III nitrides go infrared: New properties and perspectives. *J. Appl. Phys.* **2009**, *106*, 011101. [CrossRef]
2. Nakamura, S.; Pearton, S.; Fasol, G. *The Blue Laser Diode*, 2nd ed.; Springer: Berlin, Germany, 2000.

3. Su, M.; Chen, C.; Rajan, S. Prospects for the application of GaN power devices in hybrid electric vehicle drive systems. *Semicond. Sci. Technol.* **2013**, *28*, 074012. [CrossRef]

4. Kako, S.; Holmes, M.; Sergent, S.; Burger, M.; As, D.J.; Arakawa, Y. Single-photon emission from cubic GaN quantum dots. *Appl. Phys. Lett.* **2014**, *104*, 011101. [CrossRef]

5. Chou, H.T.; Goldhaber-Gordon, D.; Schmult, S.; Manfra, M.J.; Sergent, A.M.; Molnar, R.J. Single-electron transistors in GaN/AlGaN heterostructures. *Appl. Phys. Lett.* **2006**, *89*, 033104. [CrossRef]

6. D'Amico, I.; Biolatti, E.; Rossi, F.; Derinaldis, S.; Rinaldis, R.; Cingolani, R. GaN quantum dot based quantum information/computing processing. *Superlattices Microstruct.* **2002**, *31*, 117–125. [CrossRef]

7. Novikov, S.V.; Kent, A.J.; Foxon, C.T. Molecular beam epitaxy as a growth technique fpr achieving free-standing zinc-blende GaN and wurtzite $Al_xGa_{1-x}N$. *Prog. Cryst. Growth Charact. Mater.* **2017**, *63*, 25–39. [CrossRef]

8. Susanto, I.; Kan, K.Y.; Yu, I.S. Temperature effects for GaN films grown on 4H-SiC substrate with 4^0 miscutting orientation by plasma-assisted molecular beam epitaxy. *J. Alloys Compd.* **2017**, *723*, 21–29. [CrossRef]

9. Nakamura, S. GaN growth using GaN buffer layer. *Jpn. J. Appl. Phys.* **1991**, *30*, L1705–L1707. [CrossRef]

10. Sudhir, G.S.; Fujii, H.; Wong, W.S.; Kisielowski, C.; Newman, N.; Dieker, C.; Liliental-Weber, Z.; Rubin, M.D.; Weber, E.R. Pulsed laser deposition of aluminum nitride and gallium nitride thin films. *Appl. Surf. Sci.* **1998**, *127–129*, 471–476. [CrossRef]

11. Paskova, T.; Darakchieva, V.; Valcheva, E.; Paskov, P.P.; Ivanov, I.G.; Monemar, B.; Böttcher, T.; Roder, C.; Hommel, D. Hydride vapor-phase epitaxial GaN thick films for quasi-substrate applications: Strain distribution and wafer bending. *J. Electron. Mater.* **2004**, *33*, 389–394. [CrossRef]

12. Brown, J.; Wu, F.; Petroff, P.M.; Speck, J.S. GaN quantum dot density control by rf-plasma molecular beam epitaxy. *Appl. Phys. Lett.* **2004**, *84*, 690–692. [CrossRef]

13. Carlsson, N.; Seifert, W.; Petersson, A.; Castrillo, P.; Pistol, M.E.; Samuelson, L. Study of the two-dimensional–three-dimensional growth mode transition in metal-organic vapor phase epitaxy of GaInP/InP quantum-sized structures. *Appl. Phys. Lett.* **1994**, *65*, 3039. [CrossRef]

14. Koguchi, N.; Tasahashi, S.; Chikyow, T. New MBE method for InSb quantum well boxes. *J. Cryst. Growth* **1991**, *111*, 688–692. [CrossRef]

15. Wang, Z.M.; Holmes, K.; Mazur, Y.I.; Ramsey, K.A.; Salamo, G.J. Self-organization of quantum-dot pairs by high-temperature droplet epitaxy. *Nanoscale Res. Lett.* **2006**, *1*, 57–61. [CrossRef]

16. Wu, C.L.; Chou, L.J.; Gwo, S. Size- and shape-controlled GaN nanocrystals grown of Si(111) substrate by reactive epitaxy. *Appl. Phys. Lett.* **2004**, *85*, 2071–2073. [CrossRef]

17. Kondo, T.; Saitoh, K.; Yamamoto, Y.; Maruyama, T.; Naritsuka, S. Fabrication of GaN dot structures on Si substrates by droplet epitaxy. *Phys. Stat. Sol. A* **2006**, 1700–1703. [CrossRef]

18. Wang, Y.; Ozcan, A.S.; Sanborn, C.; Ludwig, K.F.; Bhattacharyya, A.; Chandrasekran, R.; Moustakas, T.D.; Zhou, L.; Smith, D.J. Real-time X-ray studies of gallium nitride nanodot formation by droplet heteroepitaxy. *J. Appl. Phys.* **2007**, *102*, 073522. [CrossRef]

19. Yu, I.S.; Chang, C.P.; Yang, C.P.; Lin, C.T.; Ma, Y.R.; Chen, C.C. Characterization and density control of GaN nanodots on Si (111) by droplet epitaxy using plasma-assisted molecular beam epitaxy. *Nanoscale Res. Lett.* **2014**, *9*, 682. [CrossRef]

20. Miller, D.A.B.; Chemla, D.S.; Damen, T.C.; Gossard, A.C.; Wiegmann, M.; Wood, T.H.; Burrus, C.A. Band-edge electroabsorption in quantum well structures: The quantum-confined stark effect. *Phys. Rev. Lett.* **1984**, *53*, 2173. [CrossRef]

21. Naritsuka, S.; Kondo, T.; Otsubo, H.; Saitoh, K.; Yamamoto, Y.; Maruyama, T. In situ annealing of GaN dot structures grown by droplet epitaxy on (111) Si substrates. *J. Cryst. Growth* **2007**, *300*, 118–122. [CrossRef]

22. Qin, H.; Luan, X.; Feng, C.; Yang, D.; Zhang, G. Mechanical, thermodynamic and electronic properties of wurtzite and zinc-blende GaN crystals. *Materials* **2017**, *10*, 1419. [CrossRef]

23. Simon, J.; Pelekanos, N.T.; Adelmann, C.; Martinez-Guerrero, E.; André, R.; Daudin, B.; Dang, L.S.; Mariette, H. Direct comparison of recombination dynamics in cubic and hexagonal GaN/AlN quantum dots. *Phys. Rev. B* **2003**, *68*, 035312. [CrossRef]

24. Martinez-Guerrero, E.; Chabuel, F.; Daudin, B.; Rouviere, J.L.; Mariette, H. Control of the morphology transition for the growth of cubic GaN/AlN nanostructures. *Appl. Phys. Lett.* **2002**, *81*, 5117. [CrossRef]

25. Gogneau, N.; Jalabert, D.; Monroy, E.; Shibata, T.; Tanaka, M.; Daudin, B. Structure of GaN quantum dots grown under modified Stranski-Krastanow conditions on AlN. *J. Appl. Phys.* **2003**, *94*, 2254–2261. [CrossRef]

26. Niu, L.; Hao, Z.; Hu, J.; Hu, Y.; Wang, L.; Luo, Y. Improving the emission efficiency of MBE-grown GaN/AlN QDs by strain control. *Nanoscale Res. Lett.* **2011**, *6*, 611. [CrossRef]

27. Schupp, T.; Meisch, T.; Neuschl, B.; Feneberg, M.; Thonke, K.; Lischka, K.; As, D.J. Droplet epitaxy of zinc-blende GaN quantum dots. *J. Cryst. Growth* **2010**, *312*, 3235–3237. [CrossRef]

28. Kawaharazuka, A.; Yoshizaki, T.; Hiratsuka, T.; Horikoshi, Y. Effect of surface Ga accumulation on the growth of GaN by molecular beam epitaxy. *Phys. Status Solidi C* **2010**, *7*, 342–346. [CrossRef]

29. Copel, M.; Reuter, M.C.; Kaxiras, E.; Tromp, R.M. Surfactants in epitaxial growth. *Phys. Rev. Lett.* **1989**, *63*, 632–635. [CrossRef]

30. Andrieu, S.; Frechard, P. What information can be obtained by RHEED applied on polycrystalline films? *Surf. Sci.* **1996**, *360*, 289–296. [CrossRef]

31. Wang, K.; Singh, J.; Pavlidis, D. Theoretical study of GaN growth: A Monte Carlo approach. *J. Appl. Phys.* **1994**, *76*, 3502–3510. [CrossRef]

32. Lymperakis, L.; Neugebauer, J. Large anisotropic adatom kinetics on nonpolar GaN surfaces: Consequences for surface morphologies and nanowire growth. *Phys. Rev. B* **2009**, *79*, 241308. [CrossRef]

33. Monu, M.; Krishna, T.C.S.; Neha, A.; Mandeep, K.; Sandeep, S.; Govind, G. Pit assisted oxygen chemisorptions on GaN surface. *Phys. Chem. Chem. Phys.* **2015**, *17*, 15201–15208. [CrossRef]

34. Chen, H.J.Y.; Su, Y.Z.; Yang, D.L.; Huang, T.W.; Yu, I.S. Effects of substrate pre-nitridation and post-nitridation processes on InN quantum dots with crystallinity by droplet epitaxy. *Surf. Coat. Technol.* **2017**, *324*, 491–497. [CrossRef]

35. CrystOrient. Available online: http://www.crystorient.com/ (accessed on 3 April 2018).

36. Romano, L.T.; Krusor, B.S.; Singh, R.; Moustakas, T.D. Structure of GaN films grown by molecular beam epitaxy on (0001) sapphire. *J. Electron. Mater.* **1997**, *26*, 285–289. [CrossRef]

Article

Influence of Solvent and Electrical Voltage on Cathode Plasma Electrolytic Deposition of Al$_2$O$_3$ Antioxidation Coatings on Ti-45Al-8.5Nb Alloys

Xu Yang [1], Zhipeng Jiang [1], Xianfei Ding [2], Guojian Hao [3], Yongfeng Liang [1] and Junpin Lin [1,*]

[1] State Key Laboratory for Advanced Metals and Materials, University of Science and Technology Beijing, Beijing 100083, China; yangxu19900808@163.com (X.Y.); 18310692178@163.com (Z.J.); liangyf@skl.ustb.edu.cn (Y.L.)

[2] Cast Titanium Alloys R&D Center, Beijing Institute of Aeronautical Materials, Beijing 100095, China; xianfeimail@gmail.com

[3] Beijing Research Institute of Mechanical & Electrical Technology, Beijing 100083, China; drhaogj@gmail.com

* Correspondence: linjunpin@ustb.edu.cn; Tel.: +86-010-6233-2192

Received: 23 March 2018; Accepted: 27 April 2018; Published: 1 May 2018

Abstract: Al$_2$O$_3$ coatings were prepared on Ti-45Al-8.5Nb alloys via cathodic plasma electrolysis deposition (CPED) in both 1.2 M Al(NO$_3$)$_3$ aqueous and ethanolic solutions. Different voltages were also applied during the deposition process to optimize coating properties. Coatings deposited in both solutions mainly consisted of γ-Al$_2$O$_3$, with some Al(OH)$_3$ found in coatings prepared in aqueous solution. Coatings prepared in ethanol solution exhibited better oxidation resistance at 900 °C as well as better substrate adhesion, which was mainly due to smaller crater sizes on coating surfaces. The deposition process was discussed in detail and the reason for the smaller craters examined. The results suggested that solution surface tension mainly influenced the average diameter of hydrogen bubbles that formed on cathode surfaces during the process. Smaller bubbles lead to both lower current densities on cathodes and smaller crater sizes on coatings.

Keywords: cathodic plasma electrolysis deposition; Al$_2$O$_3$ coating; oxidation; solution surface tension

1. Introduction

Lightweight TiAl alloys have outstanding characteristics of low density, specific strength, and elastic modulus, which yield it a key material for use in aero and aerospace applications [1–3]. At high temperature, the oxidation resistance of high Nb-containing (up to 10 at %) TiAl alloys is much higher compared with that of other TiAl alloys [4,5]. Hence, Ti-45Al-8.5Nb alloys have been chosen for use in engines working at 800–900 °C [5]. Nevertheless, non-protective Al$_2$O$_3$ + TiO$_2$ scale has been observed to grow rapidly on these alloys at high temperatures, which limits their applications [6,7]. Therefore, new or revised surface treatments are necessary to improve the alloys' oxidation resistance.

Although metal coatings can guarantee the long-term performance of TiAl alloys at high temperatures [8–10], element diffusion often appears at the interface between the coating and substrate, which, consequently, results in coating degradation [11–13]. Al$_2$O$_3$ ceramic coatings can effectively protect TiAl alloys against oxidation at high temperatures and no diffusion occurs at the interface during service [14,15], which is beneficial to both the properties of the substrate alloys and Al$_2$O$_3$ coatings [13]. Al$_2$O$_3$ coatings can be prepared by many techniques, including plasma spray (PS) [16], electron-beam physical vapor deposition (EB-PVD) [17], sol-gel [18], and electro-deposition [19,20]. However, all these techniques have their own weaknesses as well as merits. Specifically, PS and EB-PVD cannot be used for preparing sample coatings with complex shapes and, for industrial production, the sol-gel and electro-deposition should also not be used because of low efficiencies.

The process of plasma electrolysis (PE), a new surface-treatment technique, has seen wide application in ceramic coating depositions on alloys [21,22]. PE processes comprise anode and cathode plasma electrolysis, depending on the discharge electrode. As micro-arc oxidation (MAO) technique, anode plasma electrolysis had been widely used for coating metal substrates [23,24]. Nevertheless, the resulting composition and structure of coatings are subject to the substrate's influence. Because of the advantages of high deposition efficiency, low energy consumption, good coating adhesion, and environmental friendliness, cathodic plasma electrolysis deposition (CPED) had been applied on many substrate materials, including metals, alloys, silicon, and carbon [25]. Wang et al. have suggested that porous Al_2O_3 coatings can be obtained by CPED on 304 stainless steel substrate and can improve substrate corrosion resistance [26]. With H_2PtCl_6 added to the solution, Al_2O_3-Pt composite coatings have been produced and the coatings showed excellent substrate adhesion [27,28]. In addition, with polyethylene glycol (PEG) added to aqueous solutions, Al_2O_3 coatings with smaller pores have been deposited on Ni-based super alloys, which significantly improved the anti-oxidation performance of alloys [29]. According to He et al., by adding PEG or glass beads into aqueous solution, the cathode current density is significantly reduced during CPED processes [25,29]. In general, studies conducted in this field have mainly focused on coating structures and performance. Few studies have focused attention on the deposition process and the relationship between coatings properties and deposition parameters. To improve the performance of Al_2O_3 coatings prepared by CPED process, it is important to determine how solvents and voltage parameters influence coating qualities and properties.

In this work, the CPED process was adopted to prepare Al_2O_3 coatings on Ti-45Al-8.5Nb alloy substrate. $Al(NO_3)_3$ solutions with different solvents were applied as electrolyte solutions during the deposition process. The coating morphology, chemical composition, and phase components were studied and the anti-oxidation performance investigated. The relationship between solvent and power parameters and coating properties was examined in detail.

2. Materials and Methods

The substrate of cylindrical samples (6 × 50 mm) used here was Ti-45Al-8.5Nb alloy (Nb, Al, Y, and W, at 8.5, 45, 0.1, and 0.2 at %, respectively, the remainder Ti). SiC paper was used to polish samples to a grit of #400 and ultrasonic-cleaning used to clean substrates with acetone and ethanol.

The experimental apparatus for CPED (TOPWER, Yangzhou, Jiangsu, China) involved a glass beaker containing circulating cooling water to control the electrolyte temperature at 25–30 °C during the CPED process (Figure 1). A direct-current (DC) source with a voltage range of 0–400 V was connected to the graphite plate anode and the substrate material the cathode. In experiments, the anode and cathode were inserted into the electrolyte 50 mm apart. The solutions employed for preparing Al_2O_3 coatings were 1.2 M $Al(NO_3)_3$ aqueous and ethanolic solutions.

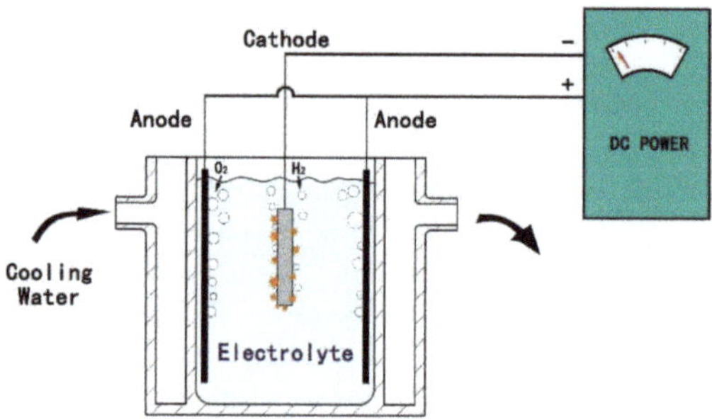

Figure 1. Experimental apparatus used for cathode plasma electrolytic deposition (CPED).

After deposition, phase constituents of deposited coatings were analyzed by X-ray diffraction (XRD) using a Rigaku Dmax-RB with Cu-K$_\alpha$ (40 kV, 40 mA, stepwise of 0.02°, and continuous scanning; Rigaku Corp., Tokyo, Japan) in the 2θ range of 20–90°. Additionally, The surface morphology of coatings were examined using scanning electron microscopy (SEM, Zeiss Supra55; Carl Zeiss Microscopy GmbH, Göttingen, Germany) in secondary electron (SE, 15 kV) imaging mode and cross sections were observed in back-scattered electron (BSE, 15 kV) mode. An energy-dispersive X-ray spectroscope (EDS, Thermo Scientific UltraDry; Thermo Fisher Scientific Inc., Pittsburgh, PA, USA) was used to analyze coating compositions. The adhesion properties of coatings were evaluated with a multi-function material surface tester (MFT-4000; Meriam, Wilmington, NC, USA). Dynamic contact angle measuring instruments and tensiometers (DCAT21; DataPhysics Instruments GmbH, Filderstadt, Germany) were applied to measure the surface tension of solutions.

Cyclic oxidation tests were conducted in a box-type resistance furnace at 900 °C in air for 100 h. During tests, samples were taken out from the furnace after 10 h, cooled to room temperature, and weighed. Then, samples were placed into the furnace again for another round of heating. The weight gained by samples during tests was measured with a 10^{-5} g-accuracy analytical balance. In a whole test period, 10 such rounds of heating were carried out to examine coating oxidation resistances.

3. Results

3.1. Current-Voltage Properties of Two Types of Solution

Current density-voltage (i_c-V) curves in aqueous and ethanolic solutions were measured (Figure 2a) and it was found that the curves were similar to those measured in other previously reported solutions [30]. Regions larger than the critical voltage A_0 in aqueous solution were suitable for the plasma process and the critical voltage in ethanolic solution termed E_0. Compared to current density (i_c) measured in aqueous solution, i_c in ethanolic solution was much lower although the critical voltage (E_0) was much larger. The reaction in ethanolic solution was also much slower than that in aqueous solution. The current density-time curves at selected voltages of 110 V (critical plasma voltage) in aqueous solution and at 180 V in ethanolic solutions are shown in Figure 2b. The cathode current density in aqueous solution changed little over the processing time. However, current density declined markedly within 60 s in ethanolic solution and the plasma intensity on the cathode surface also significantly decreased. As the stable current density in ethanolic solution was lower (~20 mA/cm^2), such that the deposition time in ethanolic solution needed to be extended to obtain coatings with identical thicknesses with that obtained in aqueous solution. The experimental parameters of different samples in both solutions are shown in Table 1. All samples were conducted in constant electrode voltage mode and the processing times for samples deposited in aqueous and ethanolic solutions were 1 and 10 min.

(a) 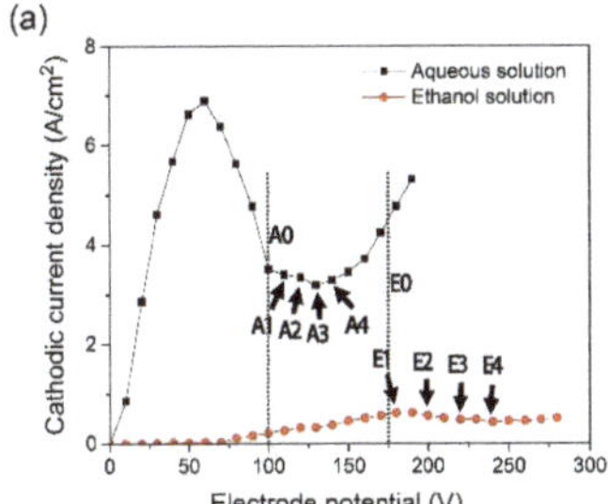
(b) 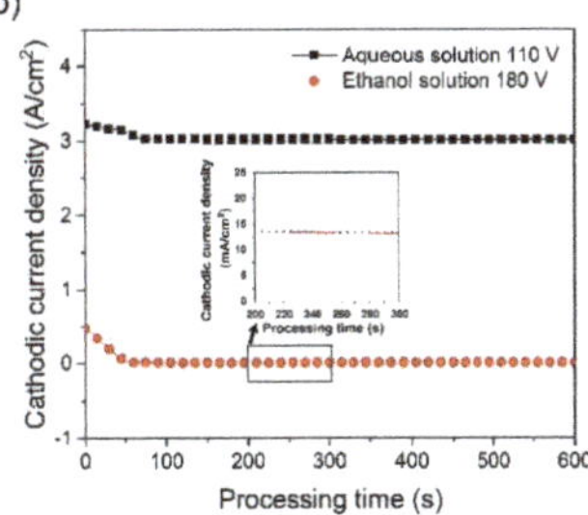

Figure 2. Typical cathodic current density versus voltage in the two solutions (**a**) and cathodic current density versus processing time (**b**).

Table 1. Sample processing parameters.

Solution	Sample	Voltage (V)	Processing Time (min)
Aqueous	A1	110	1
	A2	120	1
	A3	130	1
	A4	140	1
Ethanol	E1	180	10
	E2	200	10
	E3	220	10
	E4	240	10

3.2. Characterization of Prepared Coatings

The surface and cross-section morphologies of Al_2O_3 coatings on samples after CPED in aqueous solution at different voltages are shown in Figure 3. For samples treated at 110 V, most surfaces were covered with a thin layer of Al_2O_3 and some small craters observed, with the crater centers being plasma discharge channels on the surfaces. Far more craters were observed on the surfaces of samples treated at 120 V (Figure 3b) and a thicker Al_2O_3 coating accumulated within the same time, compared to the cross-section in Figure 3a. However, the coating structure was destroyed when the voltage was further raised to 130 V and most of the substrate surface remained exposed (Figure 3c). At 140 V, there was little coating on sample surfaces and the original surfaces damaged (Figure 3d). Some spheres of the same composition as the substrate were found on the surface after treatment with 140 V, which indicated that the substrate was partially melted by plasma during the process and then immediately cooled by the solution.

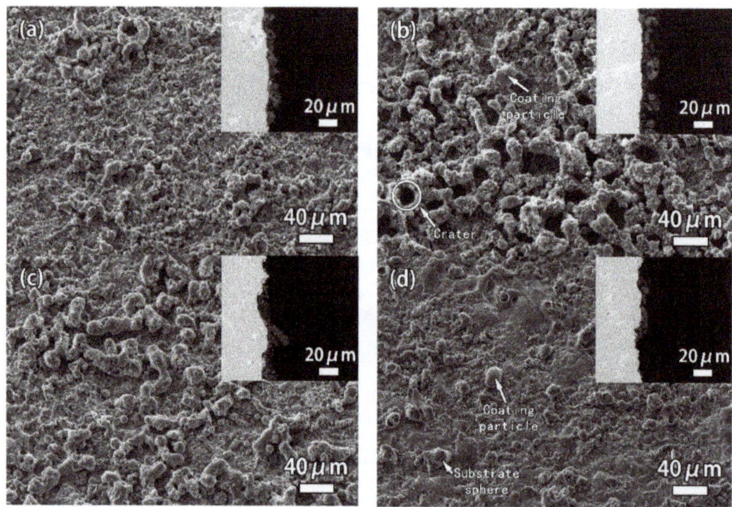

Figure 3. Typical deposition surface morphologies and cross-sections (upper-right insets) after CPED processing in aqueous solution at different voltages: (**a–d**) 110; 120; 130; and 140 V, respectively.

Surface morphologies of coatings prepared in ethanolic solution at 180–240 V for 10 min showed that coating surfaces prepared in ethanol solution (Figure 4) were clearly smoother than those obtained in aqueous solution (Figure 3). The coatings contained a large amount of small Al_2O_3 particles that particularly accumulated around small craters (Figure 4c). The crater sizes were much smaller than those observed in aqueous solution (Figure 3b) and their sizes grew with increasing electrode voltage.

Average crater diameters were ~3 μm (Figure 4a) and grew up to ~6 μm when the voltage increased to 240 V (Figure 4d). Coating thicknesses also increased significantly with increasing voltage, with the coating thickness from 180 V at ~9 μm, which reached ~27 μm at 240 V.

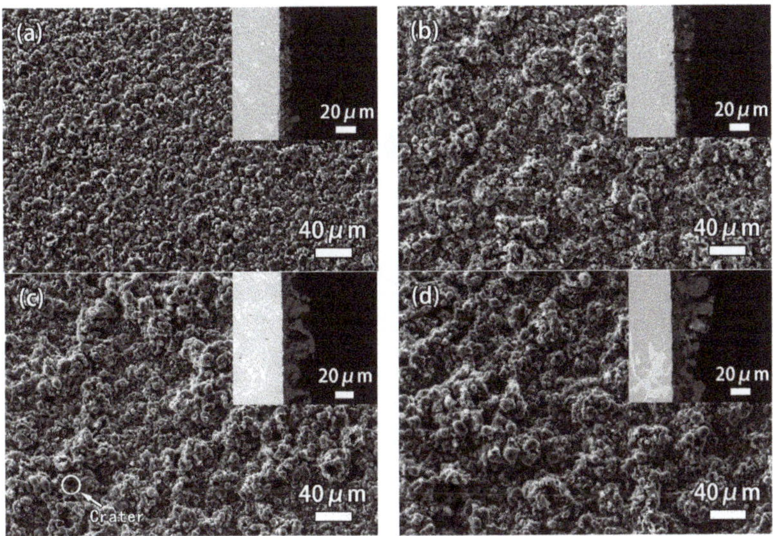

Figure 4. Typical deposition surface morphologies and cross-sections (upper-right insets) after CPED processing in ethanol solution at different voltages: (**a–d**) 180; 200; 220; and 240 V, respectively.

EDS composition results of coating surfaces showed that Al and O contents on coating surfaces increased significantly, compared with uncoated samples, which confirmed that the coatings primarily consisted of Al compounds (Table 2). The detected Ti and Nb mainly derived from the substrate, with the Ti and Nb content attributable to both the coatings' low thickness and high porosity. The results from coatings prepared at 120 V revealed that the lowest Ti component was among the A1–A4 samples (Table 2, A2), which indicated that these coatings had the maximum thickness. There was almost no Ti component in coatings produced in ethanolic solution when the voltage exceeded 200 V, which also showed that coatings were denser because the coating thickness of E2 was identical to that of A2. In addition, the atomic Al/O content was close to 2/3 in coating samples E1–E4, while the ratio was ~1/2 in samples A1–A4, which suggested that coatings fabricated in aqueous solution might contain OH^-.

Table 2. Energy-dispersive X-ray spectroscope (EDS) analysis of the surface of samples and deposition processing.

Solution	Sample	Composition (at %)			
		Al	O	Ti	Nb
-	Pure Al_2O_3	40.95	59.05	-	-
Aqueous	A1	27.84	63.2	7.84	1.12
	A2	28.52	66.01	4.46	1.01
	A3	29.13	61.64	8.25	0.97
	A4	24.71	59.21	14.54	1.54
Ethanol	E1	41.12	53.49	4.87	0.52
	E2	40.87	57.49	1.52	0.12
	E3	40.45	58.81	0.70	0.04
	E4	41.07	58.26	0.66	0.01

XRD patterns of uncoated Ti-45Al-8.5Nb alloys and deposited Al_2O_3 coatings showed that, in XRD patterns of coatings obtained in each solution (Figure 5), the phase compositions did not change with voltage. There were peaks of γ-Al_2O_3 only in coatings prepared in ethanolic solution, which was similar to coatings prepared via anodic micro-arc oxidation, which were also mainly composed of γ-Al_2O_3 [31]. However, besides the γ-Al_2O_3 component, a small amount of $Al(OH)_3$ was found in coatings deposited in aqueous solution, which agreed well with the earlier EDS analysis.

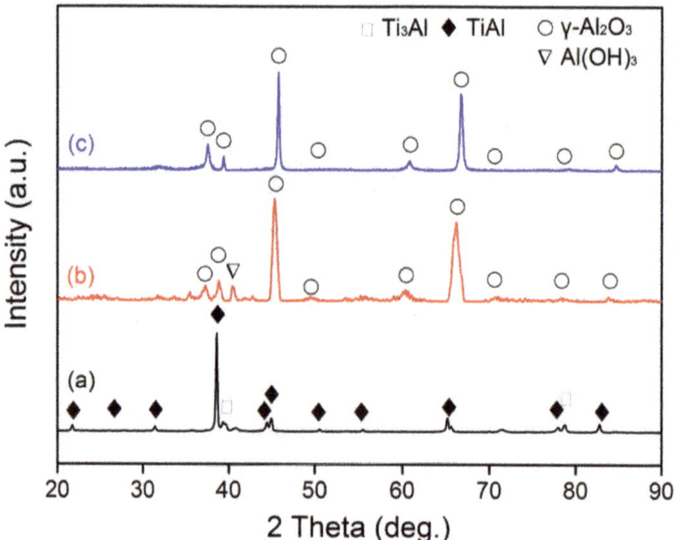

Figure 5. Sample XRD patterns of uncoated substrate (**a**) and Al_2O_3 coating obtained in aqueous (**b**) and ethanolic solutions (**c**).

3.3. Scratch Tests of Prepared Coatings

In these experiments, adhesion characteristics were examined using scratch tests. A2 and E2 were selected as examples for scratch tests because of their similar 18-μm coating thickness (Figure 6). The critical normal loads of spallation (L_S) of a coating obtained from ethanolic solution (E2) were larger than that from aqueous solution (A2). However, the critical spallation did not completely represent the adhesion property because local initial coating failure occurred at much lower loads (L_{C1}) before spallation. In addition, L_{C2} values, defined as high loads at which catastrophic failure occurred, were also different, showing how long the coating could hold and withstand further loading before catastrophic fracture. Critical loads were determined from an average of 10 readings for each scratch track and 3 scratches applied to each coating (Table 3). These results showed that E2 coating possessed a higher critical load because of the higher L_{C1}, which meant that it was more difficult to initiate a crack in the coating than in the A2 coating. The higher L_{C2} indicated that the coating prepared in ethanolic solution could also bear a larger load than that prepared in aqueous solution. Measurement of the scratch crack propagation resistance (CPR_S) was applied as a quick qualitative index of coating toughness and the results showed that coatings prepared in ethanolic solution had the higher toughness.

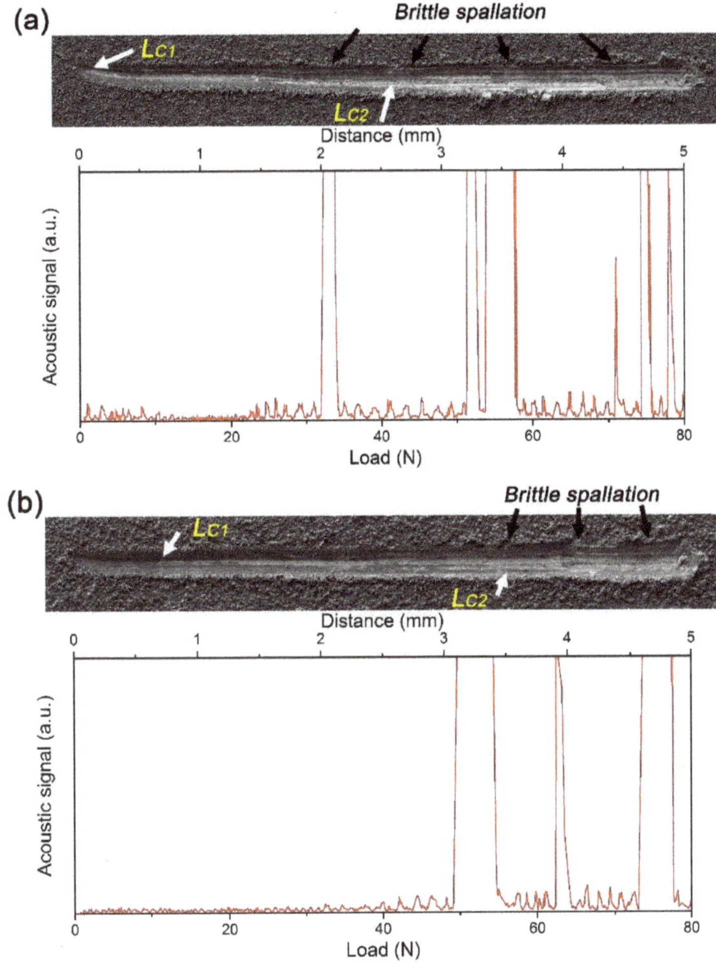

Figure 6. Results of scratch tests on Al_2O_3 coatings prepared in aqueous (**a**) and ethanolic (**b**) solutions.

Table 3. Critical loads on scratch test.

Sample	Load at Initial Failure (L_{C1}/N)	Load at Coating Spallation (L_S/N)	Load at Complete Substrate Exposure (L_{C2}/N)	CPR_S L_{C1} (L_{C2}-L_{C1})
Aqueous	1.5 ± 0.5	33.3 ± 1.5	39.5 ± 2.1	57.0
Ethanolic	8.2 ± 1.2	48.0 ± 3.7	49.8 ± 5.9	341.12

3.4. High-Temperature Cyclic Oxidation Kinetics

Oxidation kinetics curves of samples obtained during 100-h cyclic oxidation tests are shown in Figure 7. Equation (1) was used to fit the curves of sample-weight gained in cyclic oxidation tests to analyze the kinetic rule in the oxidation process

$$\Delta M = k \times t^n \tag{1}$$

where ΔM is the weight gain per unit area (mg/cm^2), n the exponential power, k the high-temperature oxidation reaction rate constant (mg/(cm$^2 \cdot$h^n)), and t the oxidation time (h). The values of n and k, after fitting the curves of mass gain for different samples, showed that sample oxidation resistances were significantly improved by Al$_2$O$_3$ coatings, compared to uncoated substrates (Table 4). Among the samples processed in aqueous solution (A1–A4), A2 had the mimimun k value, which indicated that the A2 coating had the highest high-temperature oxidation resistance because this coating had the highest thickness (Figure 3b). Sample E3 was found to exhibit the best oxidation resistance among samples processed in ethanolic solution (E1–E4). Combined with the results shown in Figure 4, sample oxidation resistance also increased with coating thickness. However, although E4's coating was thicker, its k was closer to that of E3, which was due to the larger crater sizes in the E4 coating. Furthermore, the coating formed in ethanolic solution showed higher resistance in terms of mass gains and k values of A2 and E2, because the coating thickness A2 was similar to that of E2, while A2 had higher mass gains during the oxidation process. This was also because of the larger craters sizes on the A2 coating. These results suggested that oxidation resistance had a clear, positive correlation with coating thickness and a negative correlation with porosity. Thus, a thicker coating needed to be obtained in aqueous solution to achieve the same oxidation resistance as that deposited in ethanolic solution.

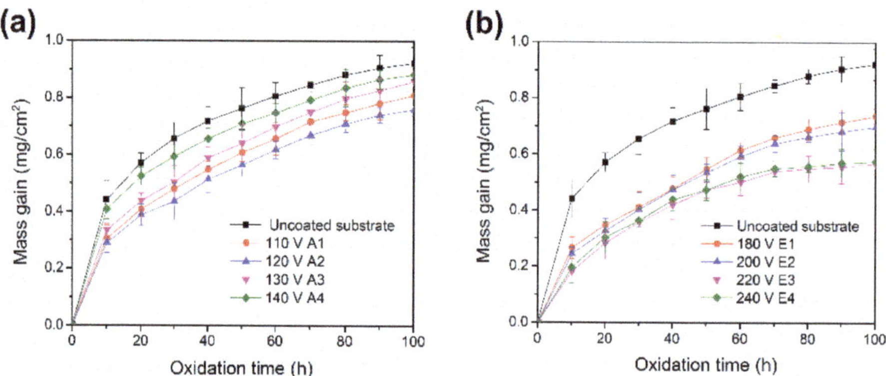

Figure 7. Cyclic oxidation kinetic curves of samples treated in aqueous (**a**) and ethanolic solution (**b**).

Table 4. The k and n values for CPED-treated samples at 900 °C after oxidation for 100 h.

Parameters	Substrate	A1	A2	A3	A4	E1	E2	E3	E4
k	0.242	0.112	0.106	0.127	0.189	0.087	0.080	0.061	0.071
n	0.291	0.431	0.430	0.415	0.337	0.468	0.481	0.498	0.472

3.5. Characterization of Samples after Oxidation

The surface morphology of samples A2 and E3 after oxidation tests showed that their morphologies changed little after oxidation (Figure 8). EDS results from different zones in Figure 8a,c are summarized in Table 5. Substrate elements were detected in the center in both craters on samples A2 and E3 (Table 5, Regions B and D), because the coating thicknesses of crater centers were smaller than in other regions. Both the crater sizes and numbers on A2 were much larger than those on E3, which led to higher mass gain in A2 during oxidation. EDS results from different positions in cross-sections in Figure 8d are also listed in Table 6. It was confirmed that the substrate was oxidized and that the oxidized scale possessed a triple-layer structure: an inner Nb-enriched, TiO$_2$-enriched, and Al$_2$O$_3$-enriched layer [5]. The oxide layer of substrate A2 was clearly thicker than that of E3, which agreed with oxidation results (Figure 7).

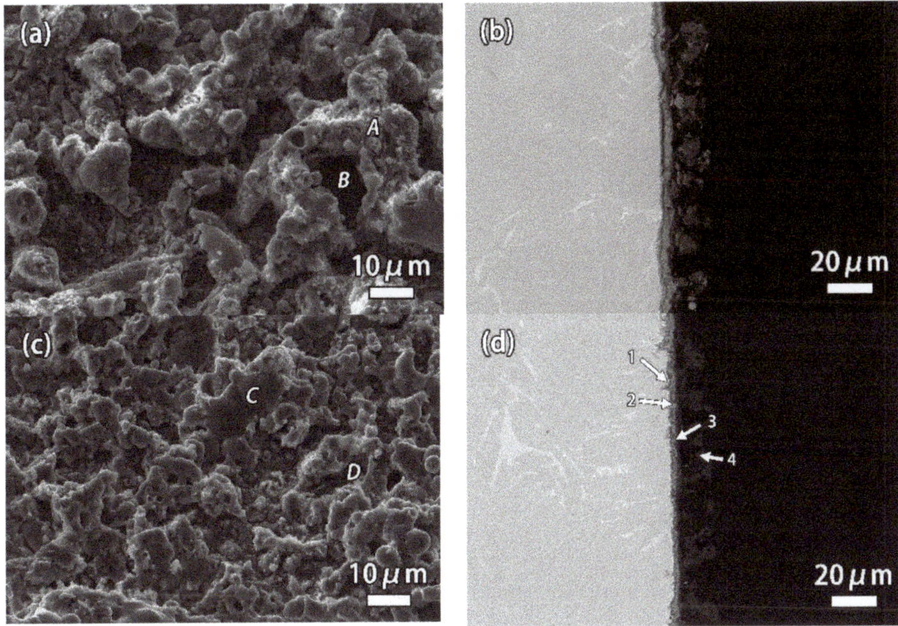

Figure 8. Surface morphologies and cross-sections of samples of A2 (**a,b**) and E3 (**c,d**) after oxidation.

Table 5. EDS analysis of different positions marked in Figure 8a,b.

Solution	Region	Composition (at %)			
		Al	O	Ti	Nb
Aqueous	A	31.12	61.11	7.14	0.62
	B	30.34	51.74	17.56	0.37
Ethanol	C	39.81	59.18	0.88	0.13
	D	31.42	54.06	14.42	0.09

Table 6. EDS analysis of different positions on the cross-section in Figure 8d.

Position	Composition (at %)			
	Al	O	Ti	Nb
1	37.08	11.93	32.91	18.07
2	9.70	55.82	26.47	8.01
3	35.12	56.49	7.82	0.57
4	47.20	51.95	0.76	0.09

XRD patterns of an uncoated sample (pattern a) and Al_2O_3-coated samples treated in aqueous solution (pattern b) and ethanolic solution (pattern c) after oxidation (Figure 9), which showed that oxidized scale of uncoated samples was mainly composed of α-Al_2O_3 and TiO_2 [5]. The α-Al_2O_3 and TiO_2 of patterns b and c also originated from substrate scales. The γ-Al_2O_3 in the original coating might have partially transformed to δ-Al_2O_3 at 900 °C, which conformed with results from a previous study [32].

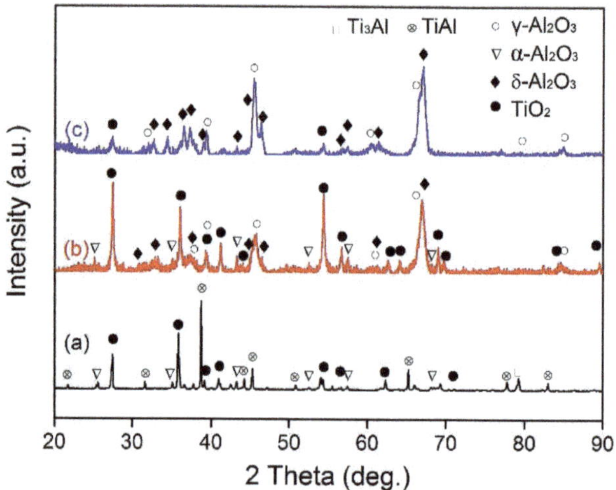

Figure 9. XRD patterns of samples after oxidation, showing phase constituents of uncoated sample (**a**) and Al_2O_3 prepared in aqueous solution (**b**) and ethanolic solution (**c**).

4. Discussion

When a sample was immersed in electrolyte as a cathode and voltage potential loaded, hydrogen reduction occurred on the cathode surface, with hydrogen bubbles appearing on the cathode. Then, the electric field strength in the bubbles increased because the covering hydrogen bubbles provided a barrier layer and cations gathered on their surfaces. The gas inside the bubbles ionized and plasma appeared when the electric field strength in the bubbles exceeded a critical value; similar descriptions have been proposed for plasma processes [30]. Some physical and chemical reactions were then promoted by high temperature around this plasma zone. Thus, here, $Al(OH)_3$ formed around the cathode surface and then transformed into Al_2O_3 [33], according to

$$Al^{3+} + 3OH^- \rightarrow Al(OH)_3 \downarrow \tag{2}$$

$$2Al(OH)_3 \overset{plasma}{\rightarrow} Al_2O_3 + 3H_2O \tag{3}$$

Hydroxide ions were mainly derived from the reduction of O_2, NO_3^-, and H_2O [34]:

$$O_2 + 2H_2O + 4e^- \rightarrow 4OH^- \tag{4}$$

$$2H_2O + 2e^- \rightarrow H_2 + 2OH^- \tag{5}$$

$$NO_3^- + H_2O + 2e^- \rightarrow NO_2^- + 2OH^- \tag{6}$$

During early deposition stages, a thin Al_2O_3 coating was deposited on the TiAl substrate, replacing the hydrogen bubbles as a barrier layer. Cathode micro-arc discharges always happened at thinner Al_2O_3-coating areas because of lower electric resistance while some Al_2O_3 coating accumulated on thinner areas. The reactions in aqueous and ethanolic solutions were similar during the deposition process but the current characteristics and coating quality differed considerably. According to the circuit law proposed by Kirchhoff, the bath voltage in electrolysis was defined by the equation [29]

$$V = \eta_a + i_c S_c R_i + \eta_c \tag{7}$$

in which, η_a is the anode overpotential, η_c that of the cathode, i_c the cathode current density, S_c the cathode surface area, and R_i the solution resistance. The current declined in ethanolic solution with increasing deposition time because the ceramic coating thickness increased such that a larger overpotential loaded on the cathode surface because of the coating electric resistance. The Al_2O_3 coating replaced the bubble film during coating growth, such that the cathode reaction area declined. The current preferentially flowed across the coating's weak regions, such that the number of tiny arcs (plasma) was also reduced. However, the current density in aqueous solution did not decrease with deposition time because the electric resistance of aqueous solution was much lower than that of the ethanolic solution. The coating's electric resistance little affected the overpotential and the bubble layer played the main role in the barrier layer for plasma formation. In addition, the crater sizes on coatings obtained from aqueous solution were larger, which further weakened its oxidation resistance. Therefore, from this point of view, ethanolic solution was more appropriate for ceramic coating deposition.

During the initial stage of the deposition process, there was no coating on the cathode surface and the current density in ethanolic solution also smaller than that in aqueous solution. Some mathematical models describing electrode surface bubble coverage and current density have been established in previous studies [35]. A hypothesis has been proposed here that constant bubble coverage of the cathode surface is required for plasma formation in different solutions. It is certain that solution surface tension reduction increases bubble coverage at a constant current density [35,36]. In other words, a lower current density is required for the same bubble coverage on the electrode when bubbles are smaller. From the surface tensions of these two solutions at 25 °C, it was clear that the ethanolic solution surface tension was much lower than that of the aqueous solution, which was the main reason why the initial current density was lower in the ethanolic solution (Table 7). Surface tensions are also known to considerably influence a bubble's departure size, which is the largest size bubble attached on the cathode surface before release, and surface tension also provides an estimate of the static bubble departure size, given by [37]

$$d = \frac{1}{2}C\left(\frac{\sigma}{g(\rho_1 - \rho_g)}\right)^{1/2} \tag{8}$$

where C represents constant coefficients, ρ_g the hydrogen gas density, ρ_1 the solution density, and σ the solution surface tension. Figure 10 shows the crater size distribution on coatings of A2 and E4, with the sizes of 300 craters per sample counted from SEM figures. A crater center was the plasma channel position on the cathode surface at which Al_2O_3 particles accumulated. Therefore, crater size was related to bubble size in the deposition process. According to Equation (8), the maximum bubble size in solution was only in relation to solution properties. Thus, the maximum crater size did not change in response to deposition voltage. The maximum crater size ratio on coatings prepared in aqueous and ethanolic solutions was ~1.67. From Equation (8), the approximate estimation of the ratio for bubble departure sizes in the two solutions was ~1.6, which was similar to the crater size ratio. Therefore, it was confirmed that a lower solution surface tension led to crater size declines on the coatings.

Table 7. Surface tension of both electrolytes.

Solution	Substrate (mN/m)
1.2 M Al(NO$_3$)$_3$ aqueous solution	70.30
1.2 M Al(NO$_3$)$_3$ ethanolic solution	26.54

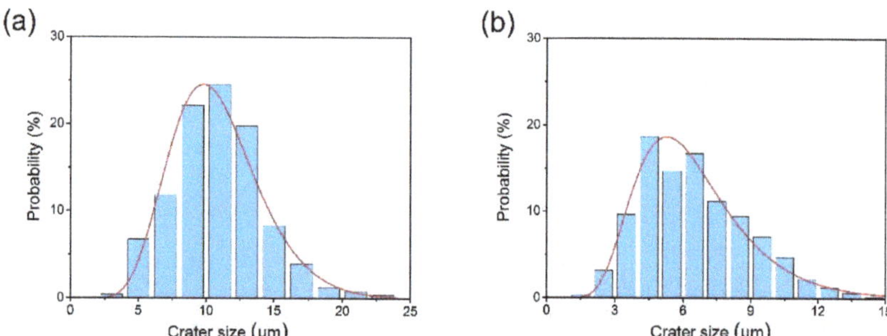

Figure 10. Size distribution of craters on Al_2O_3 coating of sample A2 (**a**) sample E4 (**b**).

Crater size also affected coating structure, as craters on a surface were held in the coating as pores during the following coating accumulation. Therefore, porosity also increased with average bubble size, significantly affecting coating properties. Discontinuity induced by pores was the main factor affecting adhesion properties and oxidation resistance. Larger porosity reduced coating strength as well as the contact surface area between coating and substrate, which then resulted in lower coating toughness in scratch tests. There was also little doubt that larger porosity was the chief factor responsible for the lower oxidation resistance. Thus, the application of ethanol as a solvent did not only improve coating quality but also would reduce energy consumption during industrial coating production. Although the processing time would be extended using ethanolic solutions, current density decreased rapidly when the ceramic coating covered most of the substrate surface. Therefore, ethanol could attain a dominant position as the preferred solvent, compared to water.

5. Conclusions

Al_2O_3 was prepared on Ti-45Al-8.5Nb alloy using the CPED process, which improved the substrate's oxidation resistance. The resulting coatings mainly consisted of γ-Al_2O_3, with coatings obtained in ethanolic solution showing smaller surface craters compared to coatings prepared in aqueous solution, which significantly affected adherence and oxidation resistance. Coatings prepared in ethanolic solution obtained higher adhesion properties with the substrate as well higher oxidation resistance. Oxidation resistance was also affected by coating thickness. The chief solution property that affected the deposition process and coating quality was solution surface tension. The application of ethanol as the solvent noticeably reduced solution surface tension, which significantly reduced bubble sizes during the CPED process and led to lower current density of CPED process and smaller coating-crater size.

Author Contributions: The work presented here was carried out in collaboration between all authors. X.Y. conceived and designed the experiments; Z.J. performed the experiments; G.H., X.D. and Y.L. analyzed the data; J.L. contributed analysis tools; and X.Y. wrote the paper.

Acknowledgments: This research was supported by the National Natural Science Foundation of China (Nos. 51671016 and 51671026).

References

1. Zhao, L.L.; Li, G.Y.; Zhang, L.Q.; Lin, J.P.; Song, X.P.; Ye, F.; Chen, G.L. Influence of Y addition on the long time oxidation behaviors of high Nb containing TiAl alloys at 900 °C. *Intermetallics* **2010**, *18*, 1586–1596. [CrossRef]

2. Froes, F.H. Production, characteristics, and commercialization of titanium aluminides. *J. Mater. Eng.* **1991**, *31*, 1235–1248. [CrossRef]

3. Yamaguchi, M.; Inui, H.; Ito, K. High-temperature structural intermetallics. *Acta Mater.* **2000**, *48*, 307–322. [CrossRef]

4. Chen, G.; Sun, Z.; Zhou, X. Oxidation and mechanical behavior of intermetallic alloys in the Ti-Nb-Al ternary system. *Mater. Sci. Eng. A* **1992**, *153*, 597–601. [CrossRef]

5. Lin, J.P.; Zhao, L.L.; Li, G.Y.; Zhang, L.Q.; Song, X.P.; Ye, F.; Chen, G.L. Effect of Nb on oxidation behavior of high Nb containing TiAl alloys. *Intermetallics* **2011**, *19*, 131–136. [CrossRef]

6. Taniguchi, S.; Uesaki, K.; Zhu, Y.C.; Zhang, H.X.; Shibata, T. Influence of niobium ion implantation on the oxidation behaviour of TiAl under thermal cycle conditions. *Mater. Sci. Eng. A* **1998**, *249*, 223–232. [CrossRef]

7. Becker, S.; Rahmel, A.; Schorr, M.; Schütze, M. Mechanism of isothermal oxidation of the intel-metallic TiAl and of TiAl alloys. *Oxid. Met.* **1992**, *38*, 425–464. [CrossRef]

8. Prasad, B.D.; Sankaran, S.N.; Wiedemann, K.E.; Glass, D.E. Platinum substitutes and two-phase-glass overlayers as a low cost alternatives to platinum aluminide coatings. *Thin Solid Films* **1996**, *345*, 255–262. [CrossRef]

9. Deodeshmukh, V.; Mu, N.; Li, B.; Gleeson, B. Hot corrosion and oxidation behavior of a novel Pt + Hf-modified γ'-Ni$_3$Al+γ-Ni-based coating. *Surf. Coat. Technol.* **2006**, *201*, 3836–3840. [CrossRef]

10. Yang, X.; Peng, X.; Wang, F. Hot corrosion of a novel electrodeposited, Ni-6Cr-7Al nanocomposite under molten (0.9Na, 0.1K)$_{(2)}$SO$_4$ at 900 degrees C. *Scripta Mater.* **2007**, *56*, 891–894. [CrossRef]

11. Miyake, M.; Tajikara, S.; Hirato, T. Fabrication of TiAl$_3$ coating on TiAl-based alloy by Al electrodeposition from dimethylsulfone bath and subsequent annealing. *Surf. Coat. Technol.* **2011**, *205*, 5141–5146. [CrossRef]

12. Wang, Y.H.; Liu, Z.G.; Ouyang, J.H.; Wang, Y.M.; Zhou, Y. Preparation and high temperature oxidation resistance of microarc oxidation ceramic coatings formed on Ti$_2$AlNb alloy. *Appl. Surf. Sci.* **2012**, *258*, 8946–8952. [CrossRef]

13. Yao, J.; He, Y.; Wang, D.; Lin, J. High-temperature oxidation resistance of (Al$_2$O$_3$–Y$_2$O$_3$)/(Y$_2$O$_3$-stabilized ZrO$_2$) laminated coating on 8Nb–TiAl alloy prepared by a novel spray pyrolysis. *Corros. Sci.* **2014**, *80*, 19–27. [CrossRef]

14. Medraj, M.; Hammond, R.; Parvez, M.A.; Drew, R.A.L.; Thompson, W.T. High temperature neutron diffraction study of the Al$_2$O$_3$–Y$_2$O$_3$ system. *J. Eur. Ceram. Soc.* **2006**, *26*, 3515–3524. [CrossRef]

15. Mah, T.I.; Keller, K.A.; Sambasivan, S.; Kerans, R.J. High-Temperature Environmental Stability of the Compounds in the Al$_2$O$_3$–Y$_2$O$_3$ System. *J. Am. Ceram. Soc.* **1997**, *80*, 874–878. [CrossRef]

16. Keyvani, A.; Saremi, M.; Sohi, M.H. Oxidation resistance of YSZ-alumina composites compared to normal YSZ TBC coatings at 1100 °C. *J. Alloys Compd.* **2011**, *509*, 8370–8377. [CrossRef]

17. Braun, R.; Fröhlich, M.; Braue, W.; Leyens, C. Oxidation behaviour of gamma titanium aluminides with EB-PVD thermal barrier coatings exposed to air at 900 °C. *Surf. Coat. Technol.* **2007**, *202*, 676–680. [CrossRef]

18. Ruhi, G.; Modi, O.P.; Sinha, A.S.K.; Singh, I.B. Effect of sintering temperatures on corrosion and wear properties of sol-gel alumina coatings on surface pre-treated mild steel. *Corros. Sci.* **2008**, *50*, 639–649. [CrossRef]

19. Kępas, A.; Grzeszczuk, M. Implications of layer-by-layer electrodeposition of polypyrrole from a solution of the same composition for ion transport in the polymer electrode. *J. Electroanal. Chem.* **2005**, *582*, 209–220. [CrossRef]

20. Cordero-Arias, L.; Boccaccini, A.R.; Virtanen, S. Electrochemical behavior of nanostructured TiO$_2$/alginate composite coating on magnesium alloy AZ91D via electrophoretic deposition. *Surf. Coat. Technol.* **2015**, *265*, 212–217. [CrossRef]

21. Gupta, P.; Tenhundfeld, G.; Daigle, E.; Ryabkov, D. Electrolytic plasma technology: Science and engineering—An overview. *Surf. Coat. Technol.* **2007**, *201*, 8746–8760. [CrossRef]

22. Yerokhin, A.; Nie, X.; Leyland, A.; Matthews, A.; Dowey, S. Plasma electrolysis for surface engineering. *Surf. Coat. Technol.* **1999**, *122*, 73–93. [CrossRef]

23. Nie, X.; Leyland, A.; Song, H.; Yerokhin, A.; Dowey, S.; Matthews, A. Thickness effects on the mechanical properties of micro-arc discharge oxide coatings on aluminium alloys. *Surf. Coat. Technol.* **1999**, *116*, 1055–1060. [CrossRef]

24. Yerokhin, A.L.; Snizhko, L.O.; Gurevina, N.L.; Leyland, A.; Pilkington, A.; Matthews, A. Matthews Discharge characterization in plasma electrolytic oxidation of aluminium. *J. Phys. D* **2003**, *36*, 2110–2120. [CrossRef]

25. Deng, S.; Wang, P.; He, Y. Influence of adding glass beads in cathode region on the kinetics of cathode plasma electrolytic depositing ZrO$_2$ coating. *Surf. Coat. Technol.* **2015**, *279*, 92–100. [CrossRef]

26. Wang, Y.; Jiang, Z.; Liu, X.; Yao, Z. Influence of treating frequency on microstructure and properties of Al_2O_3 coating on 304 stainless steel by cathodic plasma electrolytic deposition. *Appl. Surf. Sci.* **2009**, *255*, 8836–8840. [CrossRef]

27. Deng, S.; Wang, P.; He, Y.; Zhang, J. Thermal barrier coatings with Al_2O_3–Pt composite bond-coat and $La_2Zr_2O_7$–Pt top-coat prepared by cathode plasma electrolytic deposition. *Surf. Coat. Technol.* **2016**, *291*, 141–150. [CrossRef]

28. Wang, P.; He, Y.; Zhang, J. Influence of Pt particles on the porosity of Al_2O_3 coating prepared by cathode plasma electrolytic deposition. *Mater. Chem. Phys.* **2016**, *184*, 1–4. [CrossRef]

29. Wang, P.; Deng, S.; He, Y.; Liu, C.; Zhang, J. Influence of polyethylene glycol on cathode plasma electrolytic depositing Al_2O_3 anti-oxidation coatings. *Ceram. Int.* **2016**, *42*, 8229–8233. [CrossRef]

30. Meletis, E.I.; Nie, X.; Wang, F.L.; Jiang, J.C. Electrolytic plasma processing for cleaning and metal-coating of steel surfaces. *Surf. Coat. Technol.* **2002**, *150*, 246–256. [CrossRef]

31. Oliver, W.C.; Lucas, B.N.; Pharr, G.M. *Mechanical Characterization Using Indentation Experiments*; Springer: Dordrecht, The Netherlands, 1993.

32. Lee, J.; Jeon, H.; Dong, G.O.; Szanyi, J.; Kwak, J.H. Morphology-dependent phase transformation of γ-Al_2O_3. *Appl. Catal. A* **2015**, *500*, 58–68. [CrossRef]

33. Wang, X.; Liu, F.; Song, Y.; Liu, Z.; Qin, D. Structure and properties of Al_2O_3 coatings formed on NiTi alloy by cathodic plasma electrolytic deposition. *Surf. Coat. Technol.* **2016**, *285*, 128–133. [CrossRef]

34. Gal-Or, L.; Silberman, I.; Chaim, R. Electrolytic ZrO_2 Coatings. *J. Electrochem. Soc.* **1991**, *138*, 1939–1942. [CrossRef]

35. Vogt, H.; Balzer, R.J. The bubble coverage of gas-evolving electrodes in stagnant electrolytes. *Electrochim. Acta* **2005**, *50*, 2073–2079. [CrossRef]

36. Shakarji, R.A.; He, Y.; Gregory, S. The sizing of oxygen bubbles in copper electrowinning. *Hydrometallurgy* **2011**, *109*, 168–174. [CrossRef]

37. Jiang, B.; Lan, S.; Wilt, K.; Ni, J. Modeling and experimental investigation of gas film in micro-electrochemical discharge machining process. *Int. J. Mach. Tool. Manuf.* **2015**, *90*, 8–15. [CrossRef]

MDPI

St. Alban-Anlage 66

4052 Basel

Switzerland

Tel. +41 61 683 77 34

Fax +41 61 302 89 18

www.mdpi.com

Metals Editorial Office

E-mail: metals@mdpi.com

www.mdpi.com/journal/metals